INVESTIGATIONS

Stuart A. Kauffman

INVESTIGATIONS

OXFORD
UNIVERSITY PRESS

2000

OXFORD
UNIVERSITY PRESS

Oxford   New York
Athens   Auckland   Bangkok   Bogotá   Buenos Aires   Calcutta
Cape Town   Chennai   Dar es Salaam   Delhi   Florence   Hong Kong   Istanbul
Karachi   Kuala Lumpur   Madrid   Melbourne   Mexico City   Mumbai
Nairobi   Paris   São Paulo   Singapore   Taipei   Tokyo   Toronto   Warsaw

and associated companies in
Berlin   Ibadan

Published by Oxford University Press, Inc.
198 Madison Avenue, New York, New York, 10016

Oxford is a registered trademark of Oxford University Press

Library of Congress Cataloging-in-Publication Data
Kauffman, Stuart A.
Investigations / by Stuart A. Kauffman.
p. cm.
Includes bibliographical references and index.
ISBN 0–19–512104–X
1. Life—Origin—Philosophy. 2. Self-organizing systems—Philosophy. 3. Molecular evolution—
Philosophy. 4. Evolution (Biology)—Philosophy. I. Title.
QH325 .K389 2000
576.8'8'01—dc21
99–055452
0–19–512104–X

Book design and composition by Mark McGarry, Texas Type & Book Works
Set in Minion

9 8 7 6 5 4 3 2 1
Printed in the United States of America
on acid-free paper

An organized being is then not a mere machine, for that has merely *moving* power, but it possesses in itself *formative* power of a self-propagating kind which it communicates to its materials though they have it not of themselves; it organizes them, in fact, and this cannot be explained by the mere mechanical faculty of motion.

IMMANUEL KANT

Critique of Judgment

DISCARD

CONTENTS

PREFACE

Investigations is, without a doubt, the strangest and most surprising pilgrimage of my scientific life. At the outset, in December of 1994, I began an ongoing notebook. Thinking of Wittgenstein and his shattering break with logical atomism in his *Philosophical Investigations* and sensing that something vast and inarticulate waited to be uncovered, I entitled that notebook "Investigations." For the entire winter, spring, and summer of 1995 and again in the fall of that year, after a month trekking in the Everest region, I struggled sideways, backward, and sometimes forward. A rough version was published as a Santa Fe Institute preprint in September of 1996. I returned to the topic a year later and have worked at it since then. Whatever *Investigations* is—useful, as I hope, or foolish—it is not normal science.

Perhaps the most astonishing aspect of *Investigations*, both as process and the resulting book, was my puzzled realization that the way Newton, Einstein, and Bohr taught us to do science may be incomplete. You see, in following their cornerstone examples of physics, we are taught to prestate the particles, forces, laws, and initial and boundary conditions, then compute the consequences. In this enterprise, we are able to state ahead of time what the full space of possibilities is, that is, we can finitely prestate the configuration space of possibilities of the system in question. This capacity to prestate the configuration space, for example, is the central conceptual presupposition of the classical statistical mechanics of a liter of gas in its $6N$-dimensional phase space of all possible positions and momenta of the N particles of gas.

But I was, to my deep surprise, led to doubt that we can ever prestate the configuration space of a biosphere. The core issue arises with what Darwin called "preadaptations," namely, causal consequences of parts of organisms that were not of adaptive significance in the normal environment of the organism, but might come to be of adaptive significance in some new environment and end up being selected by natural selection. Thus arose hearing, lungs, flight—virtually all major and probably most or all minor adaptations. But could we say ahead of time all the odd, context-dependent causal consequences of bits and pieces of organisms that might be of selective significance in some odd environment, hence, to come into actual physical existence in the biosphere? I think not. And if not, then we cannot finitely prestate the configuration space of a biosphere.

So the biosphere, it seems, in its persistent evolution, is doing something literally incalculable, nonalgorithmic, and outside our capacity to predict, not due to quantum uncertainty alone, nor deterministic chaos alone, but for a different, equally, or more profound reason: Emergence and persistent creativity in the physical universe is real.

I think I am correct in the above statement. But then what are the implications for science and society? We have thought, in part, that the unfolding of society and culture could be brought under the sway of science. On the other hand, Sun Tzu, four centuries prior to Christ, and Clausewitz, in the early nineteenth century, in, respectively, *The Art of War* and *On War*, both stressed the totally unexpected ways of battle and the need for intuition and command genius, whatever the science lying behind strategy and tactics. Science and art—the practical getting on with it, *wissen* versus *können* in German, "know that" versus "know-how" in English, mingle in our daily lives. Yet *können*, "know-how," has no place in our science. Why?

But that takes me to the starting questions of *Investigations*. Consider a bacterium swimming upstream in a glucose gradient. We readily say that the bacterium is going to get food, that is, the bacterium is acting on its own behalf in an environment. Call a system able to act on its own behalf in an environment an "autonomous agent." All free-living cells and organisms are autonomous agents. But the bacterium is "just" a physical system. In its Kantian form, my core question became, What must a physical system be such that it can act on its own behalf?

The stunning fact is that autonomous agents do, every day, reach out and manipulate the universe on their own behalf. Yet that truth is nowhere in contemporary physics, chemistry, or even biology. So, what must a physical system be to be an autonomous agent?

I think I have found a plausible answer. I had not expected even the outlines of my answer. It may be that I have stumbled upon the proper definition of life itself. In the process, I have been led down strange pathways to a critique of the physicist's concept of work and work cycles—Maxwell's demon—toward the heart of a proper concept of organization that is not matter alone, energy alone, entropy alone, or information alone. In fact, a bacterial cell or colony is doing something that we cannot yet state clearly: The cell or colony is a "propagating organization," that is, that it literally constructs more of itself. What the cell and colony are doing has no statement in current physics or biology but constitutes that which constructs a biosphere.

Moreover, we are on the verge of the capacity to create novel molecular autonomous agents. When we do, or if we discover life on other planets and solar systems, science will enter a vast new phase in which we will create a "general biology," freed from the limitations of terrestrial biology. Such a general biology will necessarily lift physics and chemistry up to new levels as these disciplines and biology struggle to understand what general laws may govern biospheres anywhere in the cosmos. And the experimental creation of molecular autonomous agents will unleash a technological revolution equal to the computer revolution. We will create physical systems able to evolve by selection to carry out tasks we desire.

Investigations broaches four candidate laws for any biosphere. The deepest is related to the following question: Might there be a fourth law of thermodynamics for self-constructing open thermodynamic systems such as biospheres? There are good reasons to think that no such law is possible. Yet I believe there may well be such a law: Biospheres, as a secular trend, that is, over the long term, become as diverse as possible, literally expanding the diversity of what can happen next. In other words, biospheres expand their own dimensionality as rapidly, on average, as they can.

And the coconstructing behaviors of autonomous agents spill over to the economy, with surprising implications for the foundations of economics, for economic growth, and for the development of adaptive firms that coevolve in corporate ecosystems whose dynamics almost certainly express the same laws as do biological ecosystems, with small and large gales of Schumpeterian creative destruction, weeding out old species and technologies, ushering in the ever new species and technologies whose nonprestatable features are expressions of the very creativity of the universe.

And if we cannot prestate the configuration space of a biosphere, how can we prestate the configuration space of the universe? But this bears on the problem of time in general relativity and has led me, in collaboration with Lee Smolin, a quantum gravity cosmologist, to quite unexpected ideas about how the universe might select its own laws and, somewhat like a biosphere, coconstruct itself.

No, *Investigations* is not normal science. I do know what normal science is. My previous two books, *Origins of Order* and *At Home in the Universe*, are examples. The first sought to exemplify processes of self-organization in biology and to lay out the need to rebuild evolutionary theory as a marriage of two sources of order in biology—self-organization and selection. Fitness landscapes, coevolutionary avalanches, the unexpected ordered behavior of model genetic regulatory networks, the edge of chaos, models of the origin of life and morphogenesis, and bits of economics are to be found in *Origins* and in a latter-day form in *At Home*.

Nothing, however, had led me to expect even the outlines of *Investigations*. Nothing had led me to expect the answers I would struggle toward. And having completed *Investigations*, I remain profoundly puzzled by what I have said, despite the fact that I think I am correct.

I suspect that *Investigations* poises us for a new understanding of what it means and is to know and make our world together. I suspect that *Investigations* broadens the scope of science, yet will demand of us a novel synthesis of *wissen* and *können*, science and art, and ultimately, science and civilization.

It is a pleasure to thank colleagues who have read this manuscript and greatly aided its structure and increased its clarity: Phil Anderson, Sandra Blakeslee, Roumen Borrisov, Vince Darley, Bill Macready, Vladimir Makhankov, Lee Smolin, and Peter Wills. More than is usually the case, the remaining confusions and errors are mine alone. My thanks to Kirk Jensen, my editor at Oxford University Press, and the Press itself for persistent support and encouragement.

STUART KAUFFMAN
Santa Fe, New Mexico

INVESTIGATIONS

Chapter 1

PROLEGOMENON TO A
GENERAL BIOLOGY

Lecturing in Dublin, one of the twentieth century's most famous physicists set the stage of contemporary biology during the war-heavy year of 1944. Given Erwin Schrödinger's towering reputation as the discoverer of the Schrödinger equation, the fundamental formulation of quantum mechanics, his public lectures and subsequent book were bound to draw high attention. But no one, not even Schrödinger himself, was likely to have foreseen the consequences. Schrödinger's *What Is Life?* is credited with inspiring a generation of physicists and biologists to seek the fundamental character of living systems. Schrödinger brought quantum mechanics, chemistry, and the still poorly formulated concept of "information" into biology. He is the progenitor of our understanding of DNA and the genetic code. Yet as brilliant as was Schrödinger's insight, I believe he missed the center. *Investigations* seeks that center and finds, in fact, a mystery.

In my previous two books, I laid out some of the growing reasons to think that evolution was even richer than Darwin supposed. Modern evolutionary theory, based on Darwin's concept of descent with heritable variations that are sifted by natural selection to retain the adaptive changes, has come to view selection as the sole source of order in biological organisms. But the snowflake's delicate sixfold symmetry tells us that order can arise without the benefit of natural selection. *Origins of Order* and *At Home in the Universe* give good grounds to think that much of the order in organisms, from the origin of life itself to the stunning order in the development of a newborn child from a fertilized egg, does not reflect selection alone. Instead, much of the order in organisms, I believe, is self-organized and spontaneous. Self-organization mingles with natural selection in barely understood ways to yield the magnificence of our teeming biosphere. We must, therefore, expand evolutionary theory.

Yet we need something far more important than a broadened evolutionary theory. Despite any valid insights in my own two books, and despite the fine work of many others, including the brilliance manifest in the past three decades of molecular biology, the core of life itself remains shrouded from view. We know chunks of molecular machinery, metabolic pathways, means of membrane biosynthesis—we know many of the parts and many of the processes. But what makes a cell alive is still not clear to us. The center is still mysterious.

And so I began my notebook "Investigations" in December of 1994, a full half century after Schrödinger's *What Is Life?*, as an intellectual enterprise unlike any I had undertaken before. Rather bravely and thinking with some presumptuousness of Wittgenstein's famous *Philosophical Investigations*, which had shattered the philosophical tradition of logical atomism in which he had richly participated, I betook myself to my office at home in Santa Fe and grandly intoned through my fingers onto the computer's disc, "Investigations," on December 4, 1994. I sensed my long search would uncover issues that were then only dimly visible to me. I hoped the unfolding, ongoing notebook would allow me to find the themes and link them into something that was vast and new but at the time inarticulate.

Two years later, in September of 1996, I published a modestly well-organized version of *Investigations* as a Santa Fe Institute preprint, launched it onto the web, and put it aside for the time being. I found I had indeed been led into arenas that I had in no way expected, led by a swirl of ever new questions. I put the notebooks aside, but a year later I returned to the swirl, taking up again a struggle to see something that, I think, is right in front of us—always the hardest thing to see. This book is the fruit of these efforts. And this first chapter is but an introduction, in brief, to the themes that will be explained more fully in the following chapters. I would ask the reader to be patient with unfamiliar terms and concepts.

My first efforts had begun with twin questions. First, in addition to the known laws of thermodynamics, could there possibly be a fourth law of thermodynamics for open thermodynamic systems, some law that governs biospheres anywhere in the cosmos or the cosmos itself? Second, living entities—bacteria, plants, and animals—manipulate the world on their own behalf: the bacterium swimming upstream in a glucose gradient that is easily said to be going to get "dinner"; the paramecium, cilia beating like a Roman warship's oars, hot after the bacterium; we humans earning our livings. Call the bacterium, paramecium, and us humans "autonomous agents," able to act on our own behalf in an environment.

My second and core question became, What must a physical system be to be an autonomous agent? Make no mistake, we autonomous agents mutually construct our biosphere, even as we coevolve in it. Why and how this is so is a central subject of all that follows.

From the outset, there were, and remain, reasons for deep skepticism about the enterprise of *Investigations*. First, there are very strong arguments to say that there can be no general law for open thermodynamic systems. The core argument is simple to state. Any computer program is an algorithm that, given data, produces some sequence of output, finite or infinite. Computer programs can always be written in the form of a binary symbol string of 1 and 0 symbols. All possible binary symbol strings are possible computer programs. Hence, there is a countable, or denumerable, infinity of computer programs. A theorem states that for most computer programs, there is no compact description of the printout of the program. Rather, we must just unleash the program and watch it print what it prints. In short, there is no shorter description of the output of the program than that which can be obtained by running the program itself. If by the concept of a "law" we mean a compact description, ahead of time, of what the computer program will print then for any such program, there can be no law that allows us to predict what the program will actually do ahead of the actual running of the program.

The next step is simple. Any such program can be realized on a universal Turing machine such as the familiar computer. But that computer is an open nonequilibrium thermodynamic system, its openness visibly realized by the plug and power line that connects the computer to the electric power grid. Therefore, and I think this conclusion is cogent, there can be no general law for all possible nonequilibrium thermodynamic systems.

So why was I conjuring the possibility of a general law for open thermodynamic systems? Clearly, no such general law can hold for all open thermodynamic systems.

But hold a moment. It is we humans who conceived and built the intricate assembly of chips and logic gates that constitute a computer, typically we humans who program it, and we humans who contrived the entire power grid that supplies the electric power to run the computer itself. This assemblage of late-twentieth-century technology did not assemble itself. We built it.

On the other hand, no one designed and built the biosphere. The biosphere got itself constructed by the emergence and persistent coevolution of autonomous agents. If there cannot be general laws for all open thermodynamic systems, might there be general laws for thermodynamically open but self-constructing systems such as biospheres? I believe that the answer is yes. Indeed, among those candidate laws to be discussed in this book is a candidate fourth law of thermodynamics for such self-constructing systems.

To roughly state the candidate law, I suspect that biospheres maximize the average secular construction of the diversity of autonomous agents and the ways those agents can make a living to propagate further. In other words, on average, biospheres persistently increase the diversity of what can happen next. In effect, as we shall see later, biospheres may maximize the average sustained growth of their own "dimensionality."

Thus, the enterprise of *Investigations* soon began to center on the character of the autonomous agents whose coevolution constructs a biosphere. I was gradually led to a labyrinth of issues concerning the core features of autonomous agents able to manipulate the world on their own behalf. It may be that those core features capture a proper definition of life and that definition differs from the one Schrödinger found.

To state my hypothesis abruptly and without preamble, I think an autonomous agent is a self-reproducing system able to perform at least one thermodynamic work cycle. It will require most of this book to unfold the implications of this tentative definition.

Following an effort to understand what an autonomous agent might be— which, as just noted, involves the concept of work cycles—I was led to the concepts of work itself, constraints, and work as the constrained release of energy. In turn, this led to the fact that work itself is often used to construct constraints on the release of energy that then constitutes further work. So we confront a virtuous cycle: Work constructs constraints, yet constraints on the release of energy are required for work to be done. Here is the heart of a new concept of "organization" that is not covered by our concepts of matter alone, energy alone, entropy alone, or information alone. In turn, this led me to wonder about the relation between the emergence of constraints in the universe and in a biosphere, and the diversification of patterns of the constrained release of energy that alone constitute work and the use of that work to build still further constraints on the release of energy. How do biospheres construct themselves or how does the universe construct itself?

The considerations above led to the role of Maxwell's demon, one of the major places in physics where matter, energy, work, and information come together. The central point of the demon is that by making measurements on a system, the information gained can be used to extract work. I made a new distinction between measurements the demon might make that reveal features of nonequilibrium systems that can be used to extract work, and measurements he might make of the nonequilibrium system that cannot be used to extract work. How does the demon know what features to measure? And, in turn, how does work actually come to be extracted by devices that measure and detect displacements from equilibrium from which work can, in principle, be obtained? An example of such a device is a windmill pivoting to face the wind, then extracting work by the wind turning its vanes. Other examples are the rhodopsin molecule of a bacterium responding to a photon of light or a chloroplast using the constrained release of the energy of light to construct high-energy sugar molecules. How do such devices come into existence in the unfolding universe and in our biosphere? How does the vast web of constraint construction and constrained energy release used to construct yet more constraints happen into existence in the biosphere? In the universe itself? The answers appear not to be present in contemporary physics, chemistry, or biology. But a coevolving biosphere accomplishes just this coconstruction of propagating organization.

Thus, in due course, I struggled with the concept of organization itself, concluding that our concepts of entropy and its negative, Shannon's information theory (which was developed initially to quantify telephonic traffic and had been greatly extended since then) entirely miss the central issues. What is happening in a biosphere is that autonomous agents are coconstructing and propagating organizations of work, of constraint construction, and of task completion that continue to propagate and proliferate diversifying organization.

This statement is just plain true. Look out your window, burrow down a foot or so, and try to establish what all the microscopic life is busy doing and building and has done for billions of years, let alone the macroscopic ecosystem of plants, herbivores, and carnivores that is slipping, sliding, hiding, hunting, bursting with flowers and leaves outside your window. So, I think, we lack a concept of propagating organization.

Then too there is the mystery of the emergence of novel functionalities in evolution where none existed before: hearing, sight, flight, language. Whence this novelty? I was led to doubt that we could prestate the novelty. I came to doubt that we could finitely prestate all possible adaptations that might arise in a biosphere. In turn, I was led to doubt that we can prestate the "configuration space" of a biosphere.

But how strange a conclusion. In statistical mechanics, with its famous liter box of gas as an isolated thermodynamic system, we can prestate the configuration space of all possible positions and momenta of the gas particles in the box. Then Ludwig Boltzmann and Willard Gibbs taught us how to calculate macroscopic properties such as pressure and temperature as equilibrium averages over the configuration space. State the laws and the initial and boundary conditions, then calculate; Newton taught us how to do science this way. What if we cannot prestate the configuration space of a biosphere and calculate with Newton's "method of fluxions," the calculus, from initial and boundary conditions and laws? Whether we can calculate or not does not slow down the persistent evolution of novelty in the biosphere. But a biosphere is just another physical system. So what in the world is going on? Literally, what in the world is going on?

We have much to investigate. At the end, I think we will know more than at the outset. But *Investigations* is at best a mere beginning.

It is well to return to Schrödinger's brilliant insights and his attempt at a central definition of life as a well-grounded starting place. Schrödinger 's *What Is Life?* provided a surprising answer to his enquiry about the central character of life by posing a core question: What is the source of the astonishing order in organisms?

The standard—and Schrödinger argued, incorrect—answer, lay in statistical physics. If an ink drop is placed in still water in a petri dish, it will diffuse to a uniform equilibrium distribution. That uniform distribution is an average over an enormous number of atoms or molecules and is not due to the behavior of individual molecules. Any local fluctuations in ink concentration soon dissipate back to equilibrium.

Could statistical averaging be the source of order in organisms? Schrödinger based his argument on the emerging field of experimental genetics and the recent data on X-ray induction of heritable genetic mutations. Calculating the "target size" of such mutations, Schrödinger realized that a gene could comprise at most a few hundred or thousand atoms.

The sizes of statistical fluctuations familiar from statistical physics scale as the square root of the number of particles, N. Consider tossing a fair coin 10,000 times. The result will be about 50 percent heads, 50 percent tails, with a fluctuation of about 100, which is the square root of 10,000. Thus, a typical fluctuation from 50:50 heads and tails is 100/10,000 or 1 percent. Let the number of coin flips be 100 million, then the fluctuations are its square root, or 10,000. Dividing, 10,000/100,000,000 yields a typical deviation of .01 percent from 50:50.

Schrödinger reached the correct conclusion: If genes are constituted by as few as several hundred atoms, the familiar statistical fluctuations predicted by statistical mechanics would be so large that heritability would be essentially impossible. Spontaneous mutations would happen at a frequency vastly larger than observed. The source of order must lie elsewhere.

Quantum mechanics, argued Schrödinger, comes to the rescue of life. Quantum mechanics ensures that solids have rigidly ordered molecular structures. A crystal is the simplest case. But crystals are structurally dull. The atoms are arranged in a regular lattice in three dimensions. If you know the positions of all the atoms in a minimal-unit crystal, you know where all the other atoms are in the entire crystal. This overstates the case, for there can be complex defects, but the point is clear. Crystals have very regular structures, so the different parts of the crystal, in some sense, all "say" the same thing. As shown below, Schrödinger translated the idea of "saying" into the idea of "encoding." With that leap, a regular crystal cannot encode much "information." All the information is contained in the unit cell.

If solids have the order required but periodic solids such as crystals are too regular, then Schrödinger puts his bet on aperiodic solids. The stuff of the gene, he bets, is some form of aperiodic crystal. The form of the aperiodicity will contain some kind of microscopic code that somehow controls the development of the organism. The quantum character of the aperiodic solid will mean that small discrete changes, or mutations, will occur. Natural selection, operating on these small discrete changes, will select out favorable mutations, as Darwin hoped.

Fifty years later, I find Schrödinger's argument fascinating and brilliant. At once he envisioned what became, by 1953, the elucidation of the structure of DNA's aperiodic double helix by James Watson and Francis Crick, with the famously understated comment in their original paper that its structure suggests its mode of replication and its mode of encoding genetic information.

Fifty years later we know very much more. We know the human genome harbors some 80,000 to 100,000 "structural genes," each encoding the RNA that, after being transcribed from the DNA, is translated according to the genetic code to a linear sequence of amino acids, thereby constituting a protein. From Schrödinger to the establishment of the code required only about twenty years.

Beyond the brilliance of the core of molecular genetics, we understand much concerning developmental biology. Humans have about 260 different cell types: liver, nerve, muscle. Each is a different pattern of expression of the 80,000 or 100,000 genes. Since the work of François Jacob and Jacques Monod thirty-five years ago, biologists have understood that the protein transcribed from one gene might turn other genes on or off. Some vast network of regulatory interactions among genes and their products provides the mechanism that marshals the genome into the dance of development.

We have come close to Schrödinger's dream. But have we come close to answering his question, What is life? The answer almost surely is no. I am unable to say, all at once, why I believe this, but I can begin to hint at an explanation. *Investigations* is a search for an answer. I am not entirely convinced of what lies within this book; the material is too new and far too surprising to warrant conviction. Yet the pathways I have stumbled along, glimpsing what may be a terra nova, do seem to me to be worth serious presentation and serious consideration.

Quite to my astonishment, the story that will unfold here suggests a novel answer to the question, What is life? I had not expected even the outlines of an answer, and I am astonished because I have been led in such unexpected directions. One direction suggests that an answer to this question may demand a fundamental alteration in how we have done science since Newton. Life is doing something far richer than we may have dreamed, literally something incalculable. What is the place of law if, as hinted above, the variables and configuration space cannot be prespecified for a biosphere, or perhaps a universe? Yet, I think, there are laws. And if these musings be true, we must rethink science itself.

Perhaps I can point again at the outset to the central question of an autonomous agent. Consider a bacterium swimming upstream in a glucose gradient, its flagellar motor rotating. If we naively ask, "What is it doing?" we unhesitatingly answer something like, "It's going to get dinner." That is, without attributing consciousness or conscious purpose, we view the bacterium as acting on its own behalf in an environment. The bacterium is swimming upstream in order to obtain the glucose it needs. Presumably we have in mind something like the Darwinian criteria to unpack the phrase, "on its own behalf." Bacteria that do obtain glucose or its equivalent may survive with higher probability than those incapable of the flagellar motor trick, hence, be selected by natural selection.

An autonomous agent is a physical system, such as a bacterium, that can act on its own behalf in an environment. All free-living cells and organisms are clearly autonomous agents. The quite familiar, utterly astonishing feature of autonomous agents—E. coli, paramecia, yeast cells, algae, sponges, flat worms, annelids, all of us —is that we do, everyday, manipulate the universe around us. We swim, scramble, twist, build, hide, snuffle, pounce.

Yet the bacterium, the yeast cell, and we all are just physical systems. Physicists, biologists, and philosophers no longer look for a mysterious élan vital, some ethereal vital force that animates matter. Which leads immediately to the central, and confusing, question: What must a physical system be such that it can act on its own behalf in an environment? What must a physical system be such that it constitutes an autonomous agent? I will leap ahead to state now my tentative answer: A molecular autonomous agent is a self-reproducing molecular system able to carry out one or more thermodynamic work cycles.

All free-living cells are, by this definition, autonomous agents. To take a simple example, our bacterium with its flagellar motor rotating and swimming upstream for dinner is, in point of plain fact, a self-reproducing molecular system that is carrying out one or more thermodynamic work cycles. So is the paramecium chasing the bacterium, hoping for its own dinner. So is the dinoflagellate hunting the paramecium sneaking up on the bacterium. So are the flower and flatworm. So are you and I.

It will take a while to fully explore this definition. Unpacking its implications reveals much that I did not remotely anticipate. An early insight is that an autonomous agent must be displaced from thermodynamic equilibrium. Work cycles cannot occur at equilibrium. Thus, the concept of an agent is, inherently, a nonequilibrium concept. So too at the outset it is clear that this new concept of an autonomous agent is not contained in Schrödinger's answer. Schrödinger's brilliant leap to aperiodic solids encoding the organism that unleashed mid-twentieth-century biology appears to be but a glimmer of a far larger story.

Footprints of Destiny: The Birth of Astrobiology

The telltale beginnings of that larger story are beginning to be formulated. The U.S. National Aeronautics and Space Agency has had a long program in "exobiology," the search for life elsewhere in the universe. Among its well-known interests are SETI, a search for extraterrestrial life, and the Mars probes. Over the past three decades, a sustained effort has included a wealth of experiments aiming at discovering the abiotic origins of the organic molecules that are the building blocks of known living systems.

In the summer of 1997, NASA was busy attempting to formulate what it came to call "astrobiology," an attempt to understand the origin, evolution, and characteristics of life anywhere in the universe. Astrobiology does not yet exist—it is a field in the birthing process. Whatever the area comes to be called as it matures, it seems likely to be a field of spectacular success and deep importance in the coming century. A hint of the potential impact of astrobiology came in August 1997 with the tentative but excited reports of a Martian meteorite found in Antarctica that, NASA scientists announced, might have evidence of early Martian microbial life. The White House organized the single-day "Space Conference," to which I was pleased to be invited. Perhaps thirty-five scientists and scholars gathered in the Old Executive Office Building for a meeting led by Vice President Gore. The vice president began the meeting with a rather unexpected question to the group: If it should prove true that the Martian rock actually harbored fossilized microbial life, what would be the least interesting result?

The room was silent, for a moment. Then Stephen Jay Gould gave the answer many of us must have been considering: "Martian life turns out to be essentially identical to Earth life, same DNA, RNA, proteins, code." Were it so, then we would all envision life flitting from planet to planet in our solar system. It turns out that a minimum transit time for a fleck of Martian soil kicked into space to make it to earth is about fifteen thousand years. Spores can survive that long under desiccating conditions.

"And what," continued the vice president, "would be the most interesting result?" Ah, said many of us, in different voices around the room: Martian life is radically different from Earth life.

If radically different, then....

If radically different, then life must not be improbable.

If radically different, then life may be abundant among the myriad stars and solar systems, on far planets hinted at by our current astronomy.

If radically different and abundant, then we are not alone.

If radically different and abundant, then we inhabit a universe rife with the creativity to create life.

If radically different, then—thought I of my just published second book—we are at home in the universe.

If radically different, then we are on the threshold of a new biology, a "general biology" freed from the confines of our known example of Earth life.

If radically different, then a new science seeking the origins, evolution, characteristics, and laws that may govern biospheres anywhere.

A general biology awaits us. Call it astrobiology if you wish. We confront the vast new task of understanding what properties and laws, if any, may characterize biospheres anywhere in the universe. I find the prospect stunning. I will argue that the concept of an autonomous agent will be central to the enterprise of a general biology.

A personally delightful moment arose during that meeting. The vice president, it appeared, had read *At Home in the Universe*, or parts of it. In *At Home*, and also in this book, I explore a theory I believe has deep merit, one that asserts that, in complex chemical reaction systems, self-reproducing molecular systems form with high probability.

The vice president looked across the table at me and asked, "Dr. Kauffman, don't you have a theory that in complex chemical reaction systems life arises more or less spontaneously?"

"Yes."

"Well, isn't that just sensible?"

I was, of course, rather thrilled, but somewhat embarrassed. "The theory has been tested computationally, but there are no molecular experiments to support it," I answered.

"But isn't it just sensible?" the vice president persisted.

I couldn't help my response, "Mr. Vice President, I have waited a long time for such confirmation. With your permission, sir, I will use it to bludgeon my enemies."

I'm glad to say there was warm laughter around the table. Would that scientific proof were so easily obtained. Much remains to be done to test my theory.

Many of us, including Mr. Gore, while maintaining skepticism about the Mars rock itself, spoke at that meeting about the spiritual impact of the discovery of life elsewhere in the universe. The general consensus was that such a discovery, linked to the sense of membership in a creative universe, would alter how we see ourselves and our place under all, all the suns. I find it a gentle, thrilling, quiet, and transforming vision.

Molecular Diversity

We are surprisingly well poised to begin an investigation of a general biology, for such a study will surely involve the understanding of the collective behaviors of very complex chemical reaction networks. After all, all known life on earth is based on the complex webs of chemical reactions—DNA, RNA, proteins, metabolism, linked cycles of construction and destruction—that form the life cycles of cells. In the past decade we have crossed a threshold that will rival the computer revolution. We have learned to construct enormously diverse "libraries" of different DNA, RNA, proteins, and other organic molecules. Armed with such high-diversity libraries, we are in a position to begin to study the properties of complex chemical reaction networks.

To begin to understand the molecular diversity revolution, consider a crude estimate of the total organic molecular diversity of the biosphere. There are perhaps a hundred million species. Humans have about a hundred thousand structural genes, encoding that many different proteins. If all the genes within a species were identical, and all the genes in different species were at least slightly different, the biosphere would harbor about ten trillion different proteins. Within a few orders of magnitude, ten trillion will serve as an estimate of the organic molecular diversity of the natural biosphere. But the current technology of molecular diversity that generates libraries of more or less random DNA, RNA, or proteins now routinely produces a diversity of a hundred trillion molecular species in a single test tube.

In our hubris, we rival the biosphere.

The field of molecular diversity was born to help solve the problem of drug discovery. The core concept is simple. Consider a human hormone such as estrogen. Estrogen acts by binding to a specific receptor protein; think of the estrogen as a "key" and the receptor as a "lock." Now generate sixty-four million different small proteins, called peptides, say, six amino acids in length. (Since there are twenty types of amino acids, the number of possible hexamers is 20^6, hence, sixty-four million.) The sixty-four million hexamer peptides are candidate second keys, any one of which might be able to fit into the same estrogen receptor lock into which estrogen fits. If so, any such second key may be similar to the first key, estrogen, and hence is a candidate drug to mimic or modulate estrogen.

To find such an estrogen mimic, take many identical copies of the estrogen receptor, afix them to the bottom of a petri plate, and expose them simultaneously to all sixty-four million hexamers. Wash off all the peptides that do not stick to the estrogen receptor, then recover those hexamers that do stick to the estrogen receptor. Any such peptide is a second key that binds the estrogen receptor locks and, hence, is a candidate estrogen mimic.

The procedure works, and works brilliantly. By 1990, George Smith at the University of Missouri used a specific kind of virus, a filamentous phage that infects bacteria. The phage is a strand of RNA that encodes proteins. Among these proteins is the coat protein that packages the head of the phage as part of an infective phage particle. George cloned random DNA sequences encoding random hexamer peptides into one end of the phage coat protein gene. Each phage then carried a different, random DNA sequence in its coat protein gene, hence made a coat protein with a random six amino acid sequence at one end. The initial resulting "phage display" libraries had about twenty million of the sixty-four million different possible hexamer peptides.

Rather than using the estrogen receptor and seeking a peptide estrogen mimic that binds the estrogen receptor, George Smith used a monoclonal antibody molecule as the analogue of the receptor and sought a hexamer peptide that could bind the monoclonal antibody. Monoclonal antibody technology allows the generation of a large number of identical antibody molecules, hence George could use these as identical mock receptors. George found that, among the twenty million different phage, about one in a million would stick to his specific monoclonal antibody molecules. In fact, George found nineteen different hexamers binding to his monoclonal antibody. Moreover, the nineteen different hexamers differed from one another, on average, in three of the six amino acid positions. All had high affinity for his monoclonal antibody target.

These results have been of very deep importance. Phage display is now a central part of drug discovery in many pharmaceutical and biotechnology companies. The discovery of "drug leads" is being transformed from a difficult to a routine task. Not only is work being pursued using peptides but also using RNA and DNA sequences. Molecular diversity has now spread to the generation of high-diversity libraries of small organic molecules, an approach called "combinatorial chemistry." The promise is of high medical importance. As we understand better the genetic diversity of the human population, we can hope to create well-crafted molecules with increased efficacy as drugs, vaccines, enzymes, and novel molecular structures. When the capacity to craft such molecules is married, as it will be in the coming decades, to increased understanding of the genetic and cellular signaling pathways by which ontogeny is controlled, we will enter an era of "postgenomic" medicine. By learning to control gene regulation and cell signaling, we will begin to control cell proliferation, cell differentiation, and tissue regeneration to treat pathologies such as cancer, autoimmune diseases, and degenerative diseases.

But George Smith's experiments are also of immediate interest, and in surprising ways that will bear on our later discussion of autonomous agents.

George's experiments have begun to verify the concept of a "shape space" put forth by George Oster and Alan Perelson of the University of California, Berkeley, and Los Alamos National Laboratory more than a decade earlier. In turn, shape space suggests "catalytic task space." We will need both to understand autonomous agents.

Oster and Perelson had been concerned about accounting for the fact that humans can make about a hundred million different antibody molecules. Why, they wondered. They conceived of an abstract shape space with perhaps seven or eight "dimensions." Three of these dimensions would correspond to the three spatial dimensions, length, height, and width of a molecular binding site. Other dimensions might correspond to physical properties of the binding sites of molecules, such as charge, dipole moment, and hydrophobicity.

A point in shape space would represent a molecular shape. An antibody binds its shape complement, key and lock. But the precision with which an antibody can recognize its shape complement is finite. Some jiggle room is allowed. So an antibody molecule "covers" a kind of "ball" of complementary shapes in shape space. And then comes the sweet argument. If an antibody covers a ball, an actual volume, in shape space, then a finite number of balls will suffice to cover all of shape space. A reasonable analogy is that a finite number of Ping-Pong balls will fill up a bedroom.

But how big of a Ping-Pong ball in shape space is covered by one antibody? Oster and Perelson reasoned that in order for an immune system to protect an organism against disease, its antibody repertoire should cover a reasonable fraction of shape space. Newts, with about ten thousand different antibody molecules, have the minimal known antibody diversity. Perelson and Oster guessed that the newt repertoire must cover a substantial fraction, say about $1/e$—where e is the natural base for logarithms—or 37 percent of shape space. Dividing 37 percent by 10,000 gives the fractional volume of shape space covered by one antibody molecule. It follows that 100,000,000 such balls, thrown at random into shape space and allowed to overlap one another, will saturate shape space. So, 100 million antibody molecules is all we need to recognize virtually any shape of the size scale of molecular binding sites.

And therefore the concept of shape space carries surprising implications. Not surprisingly, similar molecules can have similar shapes. More surprisingly, very different molecules can have the same shape. Examples include endorphin and morphine. Endorphin is a peptide hormone. When endorphin binds the endorphin brain receptor, a euphoric state is induced. Morphine, a completely different kind of organic molecule, binds the endorphin receptor as well, with well-known consequences. Still more surprising, a finite number of different molecules, about a hundred million, can constitute a universal shape library. Thus, while there are vastly many different proteins, the number of effectively different shapes may only be on the order of a hundred million.

If one molecule binding to a second molecule can be thought of as carrying out a "binding task," then about a hundred million different molecules may constitute a universal toolbox for all molecular binding tasks. So if we can now create libraries with 100 trillion different proteins, a millionfold in excess of the universal library, we are in a position to begin to study molecular binding in earnest.

But there may also be a universal enzymatic toolbox. Enzymes catalyze, or speed up, chemical reactions. Consider a substrate molecule undergoing a reaction to a product molecule. Physical chemists think of the substrate and product molecules as lying in two potential "energy wells," like a ball at the bottom of one of two adjacent bowls. A chemical reaction requires "lifting" the substrate energetically to the top of the barrier between the bowls. Physically, the substrate's bonds are maximally strained and deformed at the top of this potential barrier. The deformed molecule is called the "transition state." According to transition state theory, an enzyme works by binding to and stabilizing the transition state molecule, thereby lowering the potential barrier of the reaction. Since the probability that a molecule acquires enough energy to hop to the top of the potential barrier is exponentially less as the barrier height increases, the stabilization of the transition state by the enzyme can speed up the reaction by many orders of magnitude.

Think of a catalytic task space, in which a point represents a catalytic task, where a catalytic task is the binding of a transition state of a reaction. Just as similar molecules can have similar shapes, so too can similar reactions have similar transition states, hence, such reactions constitute similar catalytic tasks. Just as different molecules can have the same shape, so too can different reactions have similar transition states, hence constitute the "same" catalytic task. Just as an antibody can bind to and cover a ball of similar shapes, an enzyme can bind to and cover a ball of similar catalytic tasks. Just as a finite number of balls can cover shape space, a finite number of balls can cover catalytic task space.

In short, a universal enzymatic toolbox is possible. Clues that such a toolbox is experimentally feasible come from many recent developments, including the discovery that antibody molecules, evolved to bind molecular features called epitopes, can actually act as catalysts.

Catalytic antibodies are obtained exactly as one might expect, given the concept of a catalytic task space. One would like an antibody molecule that binds the transition state of a reaction. But transition states are ephemeral. Since they last only fractions of a second, one cannot immunize with a transition state itself. Instead, one immunizes with a stable analogue of the transition shape; that is, one immunizes with a second molecule that represents the "same" catalytic task as does the transition state itself. Antibody molecules binding to this transition state analogue are tested. Typically, about one in ten antibody molecules can function as at least a weak catalyst for the corresponding reaction.

These results even allow a crude estimate of the probability that a randomly chosen antibody molecule will catalyze a randomly chosen reaction. About one antibody in a hundred thousand can bind a randomly chosen epitope. About one in ten antibodies that bind the transition state analogue act as catalysts. By this crude calculation, about one in a million antibody molecules can catalyze a given reaction.

This rough calculation is probably too high by several orders of magnitude, even for antibody molecules. Recent experiments begin to address the probability that a randomly chosen peptide or DNA or RNA sequence will catalyze a randomly chosen reaction. The answer for DNA or RNA appears to be about one in a billion to one in a trillion. If we now make libraries of a hundred trillion random DNA, RNA, and protein molecules, we may already have in hand universal enzymatic toolboxes. Virtually any reaction, on the proper molecular scale of reasonable substrates and products, probably has one or more catalysts in such a universal toolbox.

In short, among the radical implications of molecular diversity is that we already possess hundreds of millions of different molecular functions—binding, catalytic, structural, and otherwise.

In our hubris, we rival the biosphere.

In our humility, we can begin to formulate a general biology and begin to investigate the collective behaviors of hugely diverse molecular libraries. Among these collective behaviors must be life itself.

Life as an Emergent Collective Behavior
of Complex Chemical Networks

In the summer of 1996, Philip Anderson, a Nobel laureate in physics, and I accompanied Dr. Henry MacDonald, incoming director of NASA, Ames, to NASA headquarters. Our purpose was to discuss a new linked experimental and theoretical approach to the origin-of-life problem with NASA Administrator Dan Golden and his colleague, Dr. Wesley Huntress. I was excited and delighted.

As long ago as 1971, I had published my own first foray into the origin-of-life problem as a young assistant professor in the Department of Theoretical Biology at the University of Chicago. I had wondered if life must be based on template replicating nucleic acids such as DNA or RNA double helices and found myself doubting that standard assumption. Life, at its core, depends upon autocatalysis, that is, reproduction. Most catalysis in cells is carried out by protein enzymes. Might there be general laws supporting the possibility that systems of catalytic polymers such as proteins might be self-reproducing? Proteins are, as noted, linear sequences of twenty kinds of standard amino acids. Consider, then, a first copy of a protein that has the capacity to catalyze a reaction by which two fragments of a potential second copy of that same protein might be ligated to make the second copy of the whole protein. Such a protein, A, say, thirty-two amino acids long, might act on two fragments, say, fifteen amino acids and seventeen amino acids in length, and ligate the two to make a second copy of the thirty-two amino acid sequence.

But if one could imagine a molecule, A, catalyzing its own formation from its own fragments, could one not imagine two proteins, A and B, having the property that A catalyzes the formation of B by ligating B's fragments into a second copy of B, while B catalyzes the formation of A by catalyzing the ligation of A's fragments into a second copy of A? Such a little reaction system would be *collectively autocatalytic*. Neither A alone, nor B alone, would catalyze its own formation. Rather the AB system would jointly catalyze its reproduction from A and B fragments. But if A and B might achieve collective autocatalysis, might one envision a system with tens or hundreds of proteins, or peptides, that were collectively autocatalytic?

Might collective autocatalysis of proteins or similar polymers be the basic source of self-reproduction in molecular systems? Or must life be based on template replication, as envisioned by Watson and Crick, or as envisioned even earlier by Schrödinger in his aperiodic solid with its microcode? In view of the potential for a general biology, what, in fact, are the alternative bases for self-reproducing molecular systems here and anywhere in the cosmos? Which of these alternatives is more probable, here and anywhere?

By 1971 I had asked and found a preliminary answer to the following question: In a complex mixture of different proteins, where the proteins might be able to serve as candidates to ligate one another into still larger amino acid sequences, what are the chances that such a system will contain one or more collectively auto-catalytic sets of molecules? In the next chapter I will discuss the current state of theory and experiment on this issue. The best current guess is that, as the molecular diversity of a reaction system increases, a critical threshold is reached at which collectively autocatalytic, self-reproducing chemical reaction networks emerge spontaneously.

If this view is correct, and the kinetic conditions for rapid reactions can be sustained, perhaps by enclosure of such a reproducing system in a bounding membrane vesicle, also synthesized by the system, the emergence of self-reproducing molecular systems may be highly probable. No small conclusion this: Life abundant, emergent, expected. Life spattered across megaparsecs, galaxies, galactic clusters. We as members of a creative, mysteriously unfolding universe. Moreover, the hypothesis is richly testable and, as described in the next chapter, is now under the early stages of testing.

One way or another, we will discover a second life—crouched under a Mars rock, frozen in time; limpid in some pool on Titan, in some test tube in Nebraska in the next few decades. We will discover a second life, one way or another.

What monumental transformations await us, proudly postmodern, mingled with peoples on this very globe still wedded to archetypes thousands of years old.

The Strange Thing About the Theory of Evolution

We do not understand evolution. We live it with moss, fruit, fin, and quill fellows. We see it since Darwin. We have insights of forms and their formation, won from efforts since Aristotle codified the embryological investigations that over twenty-five centuries ago began with the study of deformed fetuses in sacrificial animals.

But we do not understand evolution.

"The strange thing about the theory of evolution," said one of the Huxleys (although I cannot find which one), "is that everyone thinks he understands it." How very well stated in that British fashion Americans can admire but not emulate. ("Two peoples separated by a common language," as Churchill dryly put it.)

The strange thing about the theory of evolution is that everyone thinks he understands it. How very true. It seems, of course, so simple. Finches hop around the Galapagos, occasionally migrating from island to island. Small and large beaks serve for different seeds. Beaks fitting seeds feed the young. Well-wrought beaks are selected. Mutations are the feedstock of heritable variation in a population. Populations evolve by mutation, mating, recombination, and selection to give the well-marked varieties that are, for Darwin, new species. Phylogenies bushy in the biosphere. "We're here, we're here," cry all for their typical four-million-year-stay along the four-billion-year pageant.

"We're here!"

But how?

How, in many senses. First, Darwin's theory of evolution is a theory of descent with modification. It does not yet explain the genesis of forms, but the trimmings of the forms, once they are generated. "Rather like achieving an apple tree by trimming off all the branches," said a late-nineteenth-century skeptic.

How, in the most fundamental sense: Whence life in the first place? Darwin starts with life already here. Whence life is the stuff of all later questions about whence the forms to sift.

How, in still a different sense. Darwin assumed gradualism. Most variation would be minor. Selection would sift these insensible alterations, a bit more lift, a little less drag, until the wing flew faultless in the high-hoped sky, a falcon's knot-winged, claw-latching dive to dine.

But whence the gradualism itself? It is not God given, but true, that organisms are hardly affected by most mutations. Most mutations do have little effect, some have major effects. In *Drosophila*, many mutants make small modifications in bristle number, color, shape. A few change wings to legs, eyes to antennae, heads to genitalia. Suppose that all mutations were of dramatic effect. Suppose, to take the limiting philosophical case, that all mutations were what geneticists call "lethals." Since, indeed, some mutations are lethals, one can, a priori, imagine creatures in which all mutations were lethal prior to having offspring. Might be fine creatures, too, in the absence of any mutations, these evolutionary descendants of, well, of what? And progenitors of whom? No pathway to or from these luckless ones.

Thus, evolution must somehow be crafting the very capacity of creatures to evolve. Evolution nurtures herself! But not yet in Darwin's theory, nor yet in ours.

Take another case—sex. Yes, it captures our attention, and the attention of most members of most species. Most species are sexual. But why bother? Asexuals, budding quietly wherever they bud, require only a single parent. We plumaged ones require two, a twofold loss in fitness.

Why sex? The typical answer, to which I adhere, is that sexual mating gives the opportunity for genetic recombination. In genetic recombination, the double chromosome complement sets, maternal and paternal homologues, pair up, break and recombine to yield offspring chromosomes the left half of which derives from one parental chromosome, the right half of which derives from the other parental chromosome.

Recombination is said to be a useful "search procedure" in an evolving population. Consider, a geneticist would say, two genes, each with two versions, or alleles: A and a for the first gene, B and b for the second gene. Suppose A confers a selective advantage compared to a, and B confers an advantage with respect to b. In the absence of sex, mating, and recombination, a rabbit with A and b would have to wait for a mutation to convert b to B. That might take a long time. But, with mating and recombination, a rabbit with A on the left end of a maternal chromosome, and B on the right end of the homologous paternal chromosome might experience recombination. A and B would now be on a single chromosome, hence be passed on to the offspring. Recombination, therefore, can be a lot faster than waiting for mutation to assemble the good, AB chromosome.

But it is not so obvious that recombination is a good idea after all. At the molecular level, the recombination procedure is rather like taking an airplane and a motorcycle, breaking both in half, and using spare bolts to attach the back half of the airplane to the front half of the motorcycle. The resulting contraption seems useless for any purpose.

In short, the very usefulness of recombination depends upon the gradualness that Darwin assumed. In later chapters I will discuss the concept of a "fitness landscape." The basic idea is simple. Consider a set of all possible frogs, each with a different genotype. Locate each frog in a high-dimensional "genotype space," each next to all genotypes that differ from it by a single mutation. Imagine that you can measure the fitness of each frog. Graph the fitness of each frog as a height above that position in genotype space. The resulting heights form a fitness landscape over the genotype space, much as the Alps form a mountainous landscape over part of Europe.

In the fitness landscape, image, mutation, recombination, and selection can conspire to pull evolving populations upward toward the peaks of high fitness. But not always. It is relatively easy to show that recombination is only a useful search procedure on smooth fitness landscapes. The smoothness of a fitness landscape can be defined mathematically by a correlation function giving the similarity of fitnesses, or heights, at two points on the landscape separated by a mutational distance. In the Alps, most nearby points are of similar heights, except for cliffs, but points fifty kilometers apart can be of very different heights. Fifty kilometers is beyond the correlation length of the Alps.

There is good evidence that recombination is only a useful search strategy on smooth, highly correlated landscapes, where the high peaks all cluster near one another. Recombination, half airplane–half motorcycle, is a means to look "between" two positions in a high-dimensional space. Then if both points are in the region of high peaks, looking between those two points is likely to uncover further new points of high fitness, or points on the slopes of even higher peaks. Thereafter, further mutation, recombination, and selection can bring the adapting population to successively higher peaks in the high-peaked region of the genotype space. If landscapes are very rugged and the high peaks do not cluster into smallish regions, recombination turns out to be a useless search strategy.

But most organisms are sexual. If organisms are sexual because recombination is a good search strategy, but recombination is only useful as a search strategy on certain classes of fitness landscapes, where did those fitness landscapes come from? No one knows.

The strange thing about evolution is that everyone thinks he understands it.

Somehow, evolution has brought forth the kind of smooth landscapes upon which recombination itself is a successful search strategy.

More generally, two young scientists, then at the Santa Fe Institute, proved a rather unsettling theorem. Bill Macready and David Wolpert called it the "no-free-lunch theorem." They asked an innocent question. Are there some search procedures that are "good" search procedures, no matter what the problem is? To formalize this, Bill and David considered a mathematical convenience—a set of all possible fitness landscapes. To be simple and concrete, consider a large three-dimensional room. Divide the room into very small cubic volumes, perhaps a millimeter on a side. Let the number of these small volumes in the room be large, say a trillion. Now consider all possible ways of assigning integers between one and a trillion, to these small volumes. Any such assignment can be thought of as a fitness landscape, with the integer representing the fitness of that position in the room.

Next, formalize a search procedure as a process that somehow samples M distinct volumes among the trillion in the room. A search procedure specifies how to take the M samples. An example is a random search, choosing the M boxes at random. A second procedure starts at a box and samples its neighbors, climbing uphill via neighboring boxes toward higher integers. Still another procedure picks a box, samples neighbors and picks those with lower integers, then continues.

The no-free-lunch theorem says that, averaged over all possible fitness landscape, no search procedure outperforms any other search procedure. What? Averaged over all possible fitness landscapes, you would do as well trying to find a large integer by searching randomly from an initial box for your M samples as you would climbing sensibly uphill from your initial box.

The theorem is correct. In the absence of any knowledge, or constraint, on the fitness landscape, on average, any search procedure is as good as any other.

But life uses mutation, recombination, and selection. These search procedures seem to be working quite well. Your typical bat or butterfly has managed to get itself evolved and seems a rather impressive entity. The no-free-lunch theorem brings into high relief the puzzle. If mutation, recombination, and selection only work well on certain kinds of fitness landscapes, yet most organisms are sexual, and hence use recombination, and all organisms use mutation as a search mechanism, where did these well-wrought fitness landscapes come from, such that evolution manages to produce the fancy stuff around us?

Here, I think, is how. Think of an organism's niche as a way of making a living. Call a way of making a living a "natural game." Then, of course, natural games evolve with the organisms making those livings during the past four billion years. What, then, are the "winning games"? Naturally, the winning games are the games the winning organisms play. One can almost see Darwin nod. But what games are those? What games are the games the winners play?

Ways of making a living, natural games, that are well searched out and well mastered by the evolutionary search strategies of organisms, namely, mutation and recombination, will be precisely the niches, or ways of making a living, that a diversifying and speciating population of organisms will manage to master. The ways of making a living presenting fitness landscapes that can be well searched by the procedures that organisms have in hand will be the very ways of making a living that readily come into existence. If there were a way of making a living that could not be well explored and exploited by organisms as they speciate, that way of making a living would not become populated. Good jobs, like successful jobholders, prosper.

So organisms, niches, and search procedures jointly and self-consistently co-construct one another! We make the world in which we make a living such that we can, and have, more or less mastered that evolving world as we make it. The same is true, I will argue, for an econosphere. A web of economic activities, firms, tasks, jobs, workers, skills, and learning, self-consistently came into existence in the last forty thousand years of human evolution.

The strange thing about the theory of evolution is that everyone thinks he understands it. But we do not. A biosphere, or an econosphere, self-consistently co-constructs itself according to principles we do not yet fathom.

Laws for a Biosphere

But there must be principles. Think of the Magna Carta, that cultural enterprise founded on a green meadow in England when John I was confronted by his nobles. British common law has evolved by precedent and determinations to a tangled web of more-or-less wisdom. When a judge makes a new determination, sets a new precedent, ripples of new interpretation pass to near and occasionally far reaches of the law. Were it the case that every new precedent altered the interpretation of all old judgments, the common law could not have coevolved into its rich tapestry. Conversely, if new precedents never sent out ripples, the common law could hardly evolve at all.

There must be principles of coevolutionary assembly for biospheres, economic systems, legal systems. Coevolutionary assembly must involve coevolving organizations flexible enough to change but firm enough to resist change. Edmund Burke was basically right. Might there be something deep here? Some hint of a law of coevolutionary assembly?

Perhaps. I begin with the simple example offered by Per Bak and his colleagues some years ago—Bak's "sand pile" and "self-organized criticality." The experiment requires a table and some sand. Drop the sand slowly on the table. The sand gradually piles up, fills the tabletop, piles to the rest angle of sand, then sand avalanches begin to fall to the floor.

Keep adding sand slowly to the sand pile and plot the size distribution of sand avalanches. You will obtain many small avalanches and progressively fewer large avalanches. In fact, you will achieve a characteristic size distribution called a "power law." Power law distributions are easily seen if one plots the logarithm of the number of avalanches at a given size on the y-axis, and the logarithm of the size of the avalanche on the x-axis. In the sand pile case, a straight line sloping downward to the right is obtained. The slope is the power law relation between the size and number of avalanches.

Bak and his friends called their sand pile "self-organized critical." Here, "critical" means that avalanches occur on all length scales, "self-organized" means that the system tunes itself to this critical state.

Many of us have now explored the application of Bak's ideas in models of coevolution that I will discuss shortly. With caveats that other explanations may account for the data, the general result is that something may occur that is like a theory of coevolutionary assembly that yields a self-organized critical biosphere with a power law distribution of small and large avalanches of extinction and speciation events. As we shall see, the best data now suggest that precisely such a power law distribution of extinction and speciation events has occurred over the past 650 million years of the Phanerozoic. In addition, the same body of theory predicts that most species go extinct soon after their formation, while some live a long time. The predicted species lifetime distribution is a power law. So too are the data.

Similar phenomena may occur in an econosphere. Small and large avalanches of extinction and speciation events occur in our technologies. A colleague, Brian Arthur, is fond of pointing out that when the car came in, the horse, buggy, buggy whip, saddlery, smithy, and Pony Express went out of business. The car paved the way for an oil and gas industry, paved roads, motels, fast-food restaurants, and suburbia. The Austrian economist Joseph Schumpeter wrote about this kind of turbulence in capitalist economies. These Schumpeterian gales of creative destruction appear to occur in small and large avalanches. Perhaps the avalanches arise in power laws. And, like species, most firms die young; some make it to old age—Storre, in Sweden, is over nine hundred years old. The distribution of firm lifetimes is again a power law.

Here are hints—common law, ecosystems, economic systems—that general principles govern the coevolutionary coconstruction of lives and livings, organisms and natural games, firms and economic opportunities. Perhaps such a law governs any biosphere anywhere in the cosmos.

I shall suggest other candidate laws for any biosphere in the course of *Investigations*. As autonomous agents coconstruct a biosphere, each must manage to categorize and act upon its world in its own behalf. What principles might govern that categorization and action, one might begin to wonder. I suspect that autonomous agents coevolve such that each makes the maximum diversity of reliable discriminations upon which it can act reliably as it swims, scrambles, pokes, twists, and pounces. This simple view leads to a working hypothesis: Communities of agents will coevolve to an "edge of chaos" between overrigid and overfluid behavior. The working hypothesis is richly testable today using, for example, microbial communities.

Moreover, autonomous agents forever push their way into novelty—molecular, morphological, behavioral, organizational. I will formalize this push into novelty as the mathematical concept of an "adjacent possible," persistently explored in a universe that can never, in the vastly many lifetimes of the universe, have made all possible protein sequences even once, bacterial species even once, or legal systems even once. Our universe is vastly nonrepeating; or, as the physicists say, the universe is vastly nonergodic. Perhaps there are laws that govern this nonergodic flow. I will suggest that a biosphere gates its way into the adjacent possible at just that rate at which its inhabitants can just manage to make a living, just poised so that selection sifts out useless variations slightly faster than those variations arise. We ourselves, in our biosphere, econosphere, and technosphere, gate our rate of discovery. There may be hints here too of a general law for any biosphere, a hoped-for new law for self-constructing systems of autonomous agents. Biospheres, on average, may enter their adjacent possible as rapidly as they can sustain; so too may econospheres. Then the hoped-for fourth law of thermodynamics for such self-constructing systems will be that they tend to maximize their dimensionality, the number of types of events that can happen next.

And astonishingly, we need stories. If, as I will suggest, we cannot prestate the configuration space, variables, laws, initial and boundary conditions of a biosphere, if we cannot foretell a biosphere, we can, nevertheless, tell the stories as it unfolds. Biospheres demand their Shakespeares as well as their Newtons. We will have to rethink what science is itself. And C. P. Snow's "two cultures," the humanities and science may find an unexpected, inevitable union.

Investigations leads us to new views of the biosphere as a coconstructing system. In the final chapter, I step beyond the central concern with autonomous agents to consider the universe itself. Again, there are hints of coconstruction—of the laws themselves, of the complexity of the universe, of geometry itself. The epilogue concludes with limits to reductionism in its strong form and an invocation to a constructivist science.

Chapter 2

THE ORIGINS OF LIFE

S TARTLING PROGRESS in understanding the potential routes to the origins of life on earth, and potentially anywhere in the cosmos, is now being made. There has been a "standard theory," which may still be true, and there have been several alternative views seen, but rather dimly, by the larger scholarly community. One among these may be emerging as the most plausible general picture of how life may assemble itself from simpler molecular constituents. On the other hand, no theory is yet established for the general conditions underlying the emergence of life from nonlife. Nor if such a general theory were established would that yet establish how life on Earth arose.

The Standard Model of the Origin of Life

The standard theory assumes that life must be based on template replication, more or less like that found in the DNA double helix. The reasons for this assumption are perfectly sensible. Figure 2.1 shows the familiar double helix, with its four famous nucleotide bases, adenine, thymidine, guanosine, and cytosine (A, T, G, and C). As Watson and Crick first realized, the structure of the molecule assures that A on one strand of the double helix hydrogen bonds to T on the other strand, while C hydrogen bonds to G. Thus, as they realized, the ATTCGG sequence of bases along one strand implies a corresponding TAAGCC along the opposite strand.

ATTCGGCCTTTGCCCTTAACGAT
|||||||||||||||||||||||
TAAGCCGGAAACGGGAATTGCTA

FIGURE 2.1   Schematic representation of double-stranded DNA. The four nucleotides, A, T, C, and G, along one strand each form a specific hydrogen bond pairing. Hydrogen bonded pairs are coupled with a short solid line in the figure. A pairs with T, C pairs with G. Thus, specification of the sequence of nucleotides along one strand specifies the sequence of nucleotides along the complementary strand.

Here, in the aperiodic sequence of bases, was Schrödinger's aperiodic solid and his microcode. It is central to Schrödinger's image, and more so to contemporary molecular biology, that the sequence of bases can be arbitrary. By virtue of the arbitrariness of these bases, the sequence can, in Schrödinger's apt image, say many different things, encode information. The arbitrary sequence does, in fact, carry the genetic code in which triplets of bases are codons, and sixty-one of the possible sixty-four triplet codons encode the twenty standard amino acids, while the three remaining codons specify translation "stop" signals. The specific sequence of codons, translated from a start site to the stop site, yields the messenger RNA that is then translated from start to stop into the linear sequence of amino acids constituting a protein.

Most biologists, myself included, would translate the fact that the DNA double helix supports an arbitrary sequence of bases into the statement that, thereby, the "code" can carry heritable information that, by mutation and selection, underlies the actual evolution of terrestrial life. And the same double helix suggests to all, as it did in 1953 to Watson and Crick, its mode of replication. For the sequence of bases along each strand do, in fact, specify the complementary sequence of bases along the other strand. ATTCGG does imply TAAGCC. And, in fact, a complex set of protein enzymes, including one called a DNA polymerase, helps replicate DNA by opening the double helix, allowing polymerase to slide along one strand and dutifully assembling the complementary strand from free nucleotides.

Finally, the DNA molecule is beautifully symmetric around its central axis of symmetry. Not all aperiodic solids can carry out this structural trick. Indeed, most aperiodic solids cannot come close. DNA's cousin, RNA, can form a similar double helix, as can certain other DNA-like polymers. It is the very symmetry of the molecule that allows the virtually perfect fit of A to T and C to G between the strands whatever their sequential arrangement along the strands. The symmetry allows the arbitrariness of bases to be consistent with the template replication mechanism.

It seems to most biologists that this beautiful double helix aperiodic structure is almost miraculously prefitted by chemistry and God for the task of being the master molecule of life.

If so, then the origin of life must be based on some form of a double-stranded aperiodic solid.

If life started from nonlife here or started on Mars then flung itself to Earth on some early bit of the red planet or started deep in the cosmic void and hitchhiked here on some bit of star stuff—however it started—if life started based on template replication of something like the DNA double helix, then it should be possible to find experimental conditions in which a DNA double helix, an RNA double helix, a single strand of DNA, a single strand of RNA, or some similar polymer could be caught, *en flagrante*, reproducing itself. And, per hypothesis, that Schrödinger aperiodic solid of a DNA or RNA strand had better reproduce itself in the absence of some fancy protein polymerase.

Why, on this standard model of the origin of life, must the DNA-like strand replicate in the absence of a protein polymerase of the kind that now catalyzes the sequential addition of the proper nucleotides to the newly synthesized strand as DNA or RNA replicates? Because in contemporary cells the synthesis of proteins requires all the fancy apparatus of transcription then translation of the genetic code from DNA to messenger RNA to proteins. But the transcription step requires protein enzymes. Worse still, proteins themselves are essential to the translation from messenger RNA into proteins. This occurs because specific protein enzymes "charge" specific transfer RNA molecules with the proper amino acid, then the charged transfer RNA utilizes its "anticodon" site to recognize the corresponding codon on the messenger RNA. In this way the amino acids are lined up in the right sequence along the messenger RNA, then stitched into a protein by a further RNA structure, the ribosome. So a contemporary cell uses proteins as the polymerases that replicate DNA and RNA and proteins to charge the transfer RNAs by which proteins are themselves constructed. And the cell uses the sequential structure of the bases in DNA and RNA to specify the sequences of amino acids in the proteins and uses an RNA structure, the ribosome, to link the amino acids together to form a specific protein sequence. Every step of this complex web of synthesis is catalyzed by some molecule, typically a protein, sometimes an RNA molecule.

A contemporary cell is, in point of fact, a *collectively autocatalytic whole* in which DNA, RNA, the code, proteins, and metabolism linking the synthesis of some molecular species to the breakdown of other "high-energy" molecular species all weave together and conspire to catalyze the entire set of reactions required for the whole cell to reproduce. Current life is not "nude" replicating DNA or RNA. So if life is to be based fundamentally on a simple nude replicating DNA or RNA sequence that is able also to carry genetic information, then that DNA or RNA sequence had better get on with reproduction without a protein enzyme.

Well, it has not worked yet. No one has succeeded in achieving experimental conditions in which a single-stranded DNA or RNA could line up free nucleotides, one by one, as complements to a single strand, catalyze the ligation of the free nucleotides into a second strand, melt the two strands apart, then enter another replication cycle. It just has not worked.

In fact, as Gerry Joyce, now at the Scripps Institute, first pointed out, when one thinks of the problem faced by such a DNA or RNA strand, the task is a bit daunting. Consider some arbitrary sequence along one strand, say, of RNA, where uracil substitutes for thymidine. So AUAAUCUCGGGCUUU might be such a single-stranded RNA sequence. We want it to line up the complementary nucleotides, U binding to A, G binding to C, then ligate them with the proper 3' to 5' phosphodiester bonds. But problems arise. First of all, nucleotides thermodynamically favor a 2' to 5' bond, linking the wrong atoms in two nucleotides. Second, at any such ligation step in growing the new strand, here the complement of the above (UACUAG...), the correct one of four nucleotides must be added to the growing chain and the other three candidates rejected. And this is true for any of the four possible currently terminal nucleotides in the growing chain, so our RNA sequence has to select four reactions specifically, in the proper context, out of sixteen possible. Chemically, enzymes select a reaction by binding to a transition state, as noted in chapter 1. The resulting differential velocities of the catalyzed and noncatalyzed reactions constitutes the very selectivity of one in the four reactions. Achieving high selectivity for the context-dependent four out of sixteen reactions is a tough task.

The bottom line is that, after thirty years of work by fine chemists such as Leslie Orgel at Scripps, it has not yet been possible to achieve a sequence of DNA or RNA or similar polymer able to line up a set of free nucleotides, ligate them into a complementary strand, melt the two strands apart, then cycle again, to create a self-reproducing molecular system.

Currently, efforts to achieve template self-reproduction by DNA or RNA, acting alone, are dwindling. Some focus remains on other cousins of these famous polynucleotides, and these efforts may succeed. Indeed, efforts with DNA or RNA may also succeed. But the search for self-reproducing molecular systems is now shifting in new directions.

Ribozymes, RNA molecules that can carry out catalysis, were discovered somewhat over a decade ago and startled the biological community. It had been known for nearly a century that proteins, the linear sequences of the twenty standard kinds of amino acids that fold into complex, typically compact, three-dimensional structures, can be capable of catalyzing reactions. Until ribozymes were discovered, all known enzymes were proteins. But the single-celled organism, Tetrahymena, turned out to have genes with an unexpected property. Tetrahymena turned out to have extra segments of DNA, now called "introns," that encoded extra segments of RNA introns. The introns lie between and separate "exons" in the transcribed

RNA. Prior to translating the RNA into a protein, some novel catalytic process spliced out the introns and ligated adjacent exons to create mature messenger RNA, which was then translated into a protein. It has turned out that the genes in eukaryotic cells in general, the nucleated eukaryotes, but not the bacterial prokaryotes, all have intron-exon structures. And it has turned out that the catalysts in question that catalyze the splicing out of the intron were often the intron RNA sequence itself!

So RNA molecules can catalyze reactions. But RNA molecules can also carry heritable genetic information and routinely do so in the case of many viruses whose genetic material is RNA, not DNA. At once, biologists had in hand a class of molecules, RNA, capable of storage of genetic information in exactly Schrödinger's sense of a microcode and simultaneously capable, in principle, of acting catalytically, based on the code, to carry out its instructions.

RNA might be, it has been felt, the true master molecule of life. At present, this has led to the "RNA world" hypothesis, according to which, at an early stage of life on Earth, reproducing and evolving molecular systems were composed almost entirely of RNA and small molecules, without benefit of DNA or proteins. On this view, DNA invaded later as a kind of parasite. Proteins, and coding for proteins by nucleotide triplets, also arose later, by a still unclear mechanism.

At present, there are two dramatically different experimental approaches to creating a pure RNA-world-replicating molecular system. Both are highly credible. Despite the brilliance of the first, carried out by Jack Szostak of Harvard Medical School, I will put my own bets on the second.

Szostak attracted admiration in 1990 with a spectacular paper in *Nature*, coauthored with Andrew Ellington, now at the University of Texas. Jack and Andy created biochemist's columns, glass tubes each filled with a different specific small organic molecule bound to the matrix material in the column. These columns are used to carry out molecular separation and isolation by column chromatography. They generated "libraries" of random RNA sequences, where a library had about ten trillion (or 10^{13}, or 10,000,000,000,000) different RNA sequences each about a hundred nucleotides in length. They literally poured copies of this library over each of their columns. Columns exposing a specific small molecule, say, *A*, will bind and retain any RNA sequences that happen to be able to bind to *A* or the matrix, and the column will pass through any RNA sequences that bind neither the matrix nor *A*. After washing out the nonbinding RNA sequences, Jack and Andy changed the chemical buffer to loosen the chemical bonds holding the remaining RNA sequences that did bind, and passed this modified buffer through the column to carry out, or "elute," the specific RNA sequences that had bound the organic molecule, *A*. Then these *A*-binding RNA sequences were amplified to make millions of copies of each, and the amplified selected library was passed over the same column. After a modest number of such selection-amplification cycles, Jack and

Andy had pulled out about 10^5, or a hundred thousand, different RNA sequences each able to bind to the specific molecule, A, on the column.

Jack and Andy called their binding RNA sequences "aptomers." A number of features of their results were startling and fascinating. First, the procedure works. One can find novel RNA sequences able to bind to a variety of molecular targets. Second, a diversity of different RNA aptomers are able to bind A. So there are many sequences with a shape that is complementary to the shape of A; many RNA keys fit the same lock. Third, the probability that a random RNA molecule 100 nucleotides long will bind to a randomly chosen A is the diversity of the binding RNA sequences derived from the library divided by the initial library, hence roughly 10^5 divided by 10^{13} or a probability of one in 10^8, or a hundred million. Conversely, this begins to suggest that a library with on the order of a hundred million different RNA molecules might, very roughly, have at least one RNA molecule able to bind any given target, the A molecule. So such a library with a hundred million or so RNA sequences is an experimental hint of a universal RNA toolbox, able to bind any target. More, very crudely, the human immune repertoire disposes of roughly a hundred million different antibody molecules. Returning to the fine arguments of George Oster and Alan Perelson, the human immune repertoire must be nearly a universal toolbox to recognize and bind to any molecular shape, A, B, C, and so forth. It is deeply interesting that in both cases roughly the same number, a hundred million, comes up as the requisite diversity for a universal binding toolbox.

But if binding arbitrary small molecules by RNA sequences is possible, what of catalysis by RNA molecules? How hard is it to coax an RNA molecule to catalyze a chosen reaction? Jack Szostak and David Bartel developed an elegant procedure to select RNA molecules capable of catalyzing a given reaction. They have found such molecules. At present, a crude guess at the probability that a randomly chosen RNA sequence will catalyze the specific reaction in the class of reactions they have examined is on the order of ten to a hundred-thousandfold less than the probability of merely binding. That is, among random RNA molecules, the probability of catalysis is about one in 10^{13}.

Since catalysts must bind the substrates and the products of a reaction in order to act and, more specifically, must bind the distorted molecular configuration representing the transition state, it seems likely that, if one were to preselect a library of RNA sequences that bind the substrates and products of a desired reaction, then about one in ten thousand of those binding RNA sequences would also catalyze the desired reaction. More recent results using random polypeptides suggests that finding polypeptide catalysts for a given reaction may be a thousandfold easier, closer to one in a billion rather than one in a trillion.

All this brings us to Jack Szostak's current efforts to evolve a self-reproducing RNA molecule. What Jack seeks is an RNA molecule that is able to serve as a polymerase, able to slide along and copy any arbitrary single-stranded RNA molecule by ligating the proper 3' to 5' phosphodiester bond between the terminal nucleotide in the growing new single strand being copied and the next proper nucleotide to be added. So Jack's polymerase must be capable of discriminating the proper one in four reactions for each of the four nucleotides in the extant single strand that is being copied, hence four out of sixteen reactions.

It is perfectly plausible that Jack Szostak will succeed. By hypothesis, such an RNA molecule would be able to copy any RNA molecule, hence able to copy itself. Jack will have created a self-reproducing RNA molecule.

My own view of such a tour de force is that it will be of high interest, but not central to a plausible origin-of-life theory. I take this stance for these reasons: Such a molecule seems very sophisticated. The time frame for such a molecule to arise haphazardly might be very long, perhaps too long for the hundred million years between the cooling of the crust and the emergence of life. But this worry is not too convincing. Lots of molecular experiments can have occurred in a hundred-million-year interval.

Of more concern to me is whether such a molecular system could evolve, even if it arose. Any such enzyme makes mistakes, catalyzes the wrong reaction. Suppose Jack Szostak's molecule made a copying mistake when copying a copy of itself. The resulting mutant molecule would presumably be even more error prone and would act on perfect and mutated copies of the initial perfect Szostak molecule to make a cloud of increasingly mutated ribozymes. These in turn would be even more error prone and act back on the perfect and only slightly mutated Szostak molecules to make even more mutated molecules. One might get what Leslie Orgel first called a runaway error catastrophe. The molecular system would tend to mutate itself and its progeny faster than natural selection could rectify the errors, and all information in the molecular system would degrade. The system would destroy itself.

There is a further concern. Living cells seem, ineluctably, to be organized "wholes." A cell is not a single kind of molecule replicating itself; it is, rather, a rich tapestry of molecular happenings by which the whole that is the cell propagates rough reproductions of itself. There is, after all, all of metabolism, the busy building of cell membranes and organelles. There is the business of DNA to RNA to protein, where the code itself is mediated by the very aminoacyl transferase charging enzymes that load the proper transfer RNA molecules with the right amino acids to translate the code that creates, among other things, the aminoacyl transferase enzymes themselves. There is the swishing of energy flowing in and through the multiplexed highways and byways, labyrinthine in their details, linking the breakdown of high energy sources to the synthesis of products that require the addition of free energy.

A living cell is, by inspection, and as already noted, a collectively autocatalytic system. No molecular species alone makes copies of itself. What is this whole? And is the holism requisite? What is the need for the molecular network's intricate weaving that appears the very soul of a cell?

In any theory of nude replicating genes, such as the standard view of "a" replicating polynucleotide, be it without enzymes or a Szostakian RNA polymerase
mumbling merrily to itself, "AAUGGCCAAUCCCC... ," the virtue is the apparent
simplicity of the initiation of life. Let the nude replicating molecule itself exist and
away spins a biosphere.

But that leaves unanswered where the holistic web of a cell comes from and,
more critically, whether the web is essential. You see, the simplest free-living systems, pleuromonia (PPLO), a simplified bacterial species that haunts the lungs of
sheep, still has a membrane, DNA, a code, perhaps three hundred assorted genes,
the machinery of transcription and translation, a metabolism, a linking of the
swishes of energy in and through.

A virtue of a nude-gene theory is that life starts simple. A vice is that it provides
no answer to the question, Why are free-living cells of an apparent minimal complexity? I suspect the minimal complexity is real. To assemble a sufficient diversity
of molecular functions that happen to widget together to compose a minimal critter capable of reproduction and evolution to higher complexity may well require a
minimal complexity. To be capable of evolving from such a progenote to the ever
enriching complexity of a biosphere may need a requisite initial variety of functionality. Indeed, years ago mathematician John von Neumann also thought that
some minimal complexity might be needed to create a system able to reproduce
and enrich that complexity.

I will return to these themes later, for a strength of the theory of collectively autocatalytic sets that I will describe is that it naturally leads to the expectation of an
ineluctable holism of a minimal complexity.

Experimental Autocatalytic Sets

Self-reproducing molecular systems have recently been achieved experimentally.
German biochemist Gunter von Kiedrowski, at the University of Freiburg in
Breisgau, created the molecular self-reproducing system several years ago. Gunter
reasoned that a single-stranded hexamer of DNA oriented 3' to 5', namely
3'CCGCGG5', might be able to bind, hence line up adjacent to one another the two
trimers that were its complements, '5GGC3' and 5'GCC3' (Figure 2.2). Once the
hexamer had bound the two trimers adjacent to one another, von Kiedrowski hoped,
then the proximity of the two trimers might help hasten a reaction binding the two
trimers together by a proper 3' to 5' phosophodiester bond. And, if so, then the newly
ligated oriented hexamer, 5'GGCGCC3', would be identical to the initial hexamer,
3'CCGCGG5' (both are the identical sequence if read in the 3' to 5' direction).

It worked! Shortly thereafter, Leslie Orgel and his colleagues created an even
simpler case, a single-stranded DNA tetramer, 3'CGCG5', that lines up two dimers,

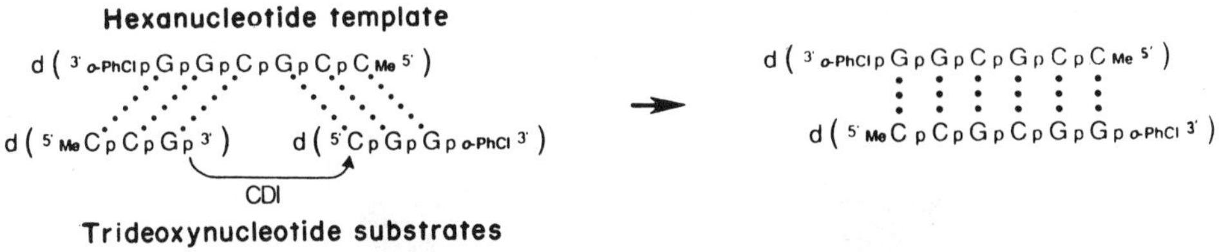

Hexanucleotide template

d (3' o-PhClp G p G p C p G p C p C Me 5')

d (5' Me C p C p G p 3')   d (5' C p G p G p o-PhCl 3')

CDI

Trideoxynucleotide substrates

→

d (3' o-PhClp G p G p C p G p C p C Me 5')

d (5' Me C p C p G p C p G p G p o-PhCl 3')

FIGURE 2.2 Self-replication of a single-stranded DNA hexamer, 3'GpGpCpGpCpCp5', through ligation catalyzed by the hexamer binding to two single stranded DNA trimers, which when ligated constitute a second copy of the initial hexamer. DNA molecules are oriented 3' to 5'. The "p" refers to the phosophate backbone linking adjacent nucleotides by phosophodiester bonds.

5'GC3' and 5'GC3', next to one another, then catalyzes the ligation of the two dimers into a second copy of itself.

In each of these cases, one contemplates a single molecule that is autocatalytic—the hexamer catalyzes the reaction that ligates two trimers into a copy of itself; the tetramer catalyzes the reaction that ligates the two dimers into a copy of itself. Naming the autocatalytic molecule A, one can say that A catalyzes its formation by ligating two fragments of A into a second copy of A.

But might it be possible for two different molecules, A and B, to catalyze one another's formation? What if A catalyzes the formation of B from B's fragments, and B catalyzes the formation of A from A's fragments? As shown in Figure 2.3, such a catalytic system is collectively autocatalytic. In a collectively autocatalytic system, no molecule catalyzes its own formation, but the set of molecules as a whole catalyzes its own reproduction from input molecular species. In this simplest case, A and B are collectively autocatalytic, and the input molecular species are the two fragments of A and the two fragments of B.

Well, it's been done, also by Gunter von Kiedrowski, who created two single-stranded DNA hexamers, A and B, where A catalyzed the ligation of two fragments to form B, while B catalyzed the ligation of two fragments of A to form A.

There are two enormous messages in these initial experiments: First, reproducing molecular systems do not need to be based on the sequential addition of single nucleotides, A,U,C,C,G,G,C, as the template complements of an initial single-stranded RNA molecule, UAGGCCG. Instead, in Gunter's initial experiment a DNA hexamer ligates two DNA trimers held adjacent to one another. There is no sequential addition of nucleotides.

Second, collective autocatalysis, A and B jointly catalyzing one another's formation, is possible. But if two molecular species, A and B, can be collectively

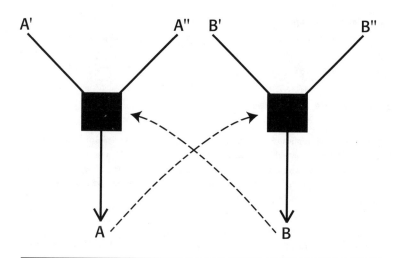

FIGURE 2.3 Hypothetical collective autocatalytic set. Polymer A, which might be a polynucleotide or short protein, is formed by ligation of two subfragments, A' and A", which when ligated end to end constitute a second copy of polymer A. Similarly, polymer B is formed by ligation of two subfragments, B' and B", which when ligated end to end form a second copy of B. Lines leading from A' and A" to the small box and the arrow from box to A represents the ligation reaction linking A' and A" to a second copy of A; similarly for the lines connecting B' and B" to B. Dashed lines from A to the B reaction box and from B to the A reaction box represent catalysis of the corresponding reaction. Thus, A catalyzes the ligation of B' and B" to form B, while B catalyzes the ligation of A' and A" to form A. The entire system is collectively autocatalytic.

autocatalytic, why not three molecular species, or ten, or a hundred, or a thousand, or a, well, a cell? Why not? Indeed, a cell is collectively autocatalytic.

But then life need not be based on template replication at all, not at all. Life can be based on far deeper principles of catalysis and what might be called "catalytic closure"—in an autocatalytic set, all the molecules whose formation must be catalyzed find molecular species within the set that catalyzes the reactions forming each of those molecules. All the "catalytic tasks" get done such that the set is collectively autocatalytic. This holism is not mystical; it is instead an objective, observable property of a collectively autocatalytic set of molecules.

The radical new view of life that I adhere to is that life is based on collectively autocatalytic sets of molecules, not on template replication per se. And more, as I shall suggest below, the emergence of collectively autocatalytic sets of molecules is not improbable but becomes almost inevitable in sufficiently diverse chemical reaction networks.

And what about the preeminence of polynucleotides? Is DNA, is RNA, essential to self-reproduction? Or might other classes of polymers be capable of self-reproduction? What of proteins, those ubiquitous catalysts? Since Watson and Crick discovered the template symmetry of double-stranded DNA, everyone has seen how the molecule might copy itself. But proteins? Proteins fold into compact three-dimensional structures like hemoglobin. How could a mechanism copy that structure?

Well, it's hard if the aim is to copy the structure of hemoglobin in one go, but what about ligating subsequences of hemoglobin together to build up the entire sequence from its fragments? What about the conceptual possibility of a collectively autocatalytic set based entirely on proteins catalyzing one another's formation by some web of ligation reactions?

Reza Ghadiri and his colleagues have made the initial, startling discovery. These fine chemists at Scripps published in the August 1996 issue of *Nature* the first example of a self-reproducing protein. Figure 2.4 shows Ghadiri's example, a 32-amino-acid sequence, let's call it "A," that catalyzes the formation of a copy of itself by lining up and ligating two fragments of itself. The 32-amino-acid sequence folds into an alpha-helix, and the helix in turn folds back on itself to

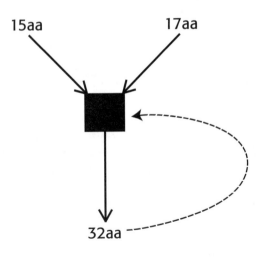

FIGURE 2.4 Ghadiri's autocatalytic 32-amino-acid-long peptide, which catalyzes the formation of additional copies of itself from 15-amino-acid-long, and 17-amino-acid-long subfragments. This is the first example of an autocatalytic peptide.

create a structure called a "coiled coil." Ghadiri and colleagues reasoned that because the alpha-helix folds back on itself, and therefore binds to itself, the same sequence would be able to bind two fragments of itself. To help supply energy to help drive the formation of the peptide bond between the adjacent fragments, Ghadiri used a chemical trick, making one fragment electrophilic (E), the other nucleophilic (N). Ghadiri and colleagues called the 32-amino-acid peptide the "template." (T). Then T lines up E and N adjacent to itself and catalyzes the ligation of E and N to create a second copy of T. The experiment worked brilliantly. Reza and his friends demonstrated that life might be based on proteins, even proteins alone.

In subsequent work, Ghadiri and his colleagues have made a "soup" of similar T, E, and N peptides, where a specific template, T_1, might not only catalyze the ligation of E_1 and N_1 to form a second copy of T_1, but T_1 might also catalyze the ligation of E_2 and N_2 to form T_2. In turn, T_2 might act not only on E_2 and N_2, but also on N_1 and E_1, or other combinations of E and N fragments.

In short, among the possibilities already demonstrated are modestly complex reaction networks of autocatalytic and cross-catalytic peptides. For example, A might catalyze its own formation as well as the formation of B, while B catalyzes its own formation and the formation of A as well, in a catalytic structure named a "hypercycle" by Nobel laureate Manfred Eigen and Peter Schuster in 1977. Ghadiri and colleagues have published the first example of a peptide hypercycle. A collectively autocatalytic network where A and B mutually catalyze one another's formation, but neither A nor B catalyzes its own formation directly, has not yet been achieved, but presumably will be in the near future. After all, Gunter von Kiedrowski has already made a collectively autocatalytic AB system with single-stranded DNA hexamers.

The Ghadiri experiments open the door to work on self-reproducing molecular systems in complex chemical reaction networks where the substrates and products are all peptides. The field of molecular diversity, generating trillions of more or less random DNA, RNA, and protein sequences, means that we can create complex reaction networks at will. Since DNA, RNA, and proteins can all bind to, and presumably also catalyze, reactions concerning the other classes of polymers, there is no reason not to seek autocatalytic and collectively autocatalytic systems of DNA, RNA, and protein species simultaneously.

Nor is the search for self-reproduction limited to linear polymers. Julius Rebek at Scripps has created a self-reproducing molecular system based on heterodimerization of complex organic molecules, calixarene ureas.

So, in the American vernacular, self-reproducing molecular systems are a done deal.

Now what?

Now much.

Now the first real hints of a general biology and a broad basis of life in the universe.

Now new concepts about the origin of life on Earth.

Now the first hints of a new technology based on self-reproducing, evolvable molecular systems.

Now the hard, hard push to explore a terra nova.

And perhaps first in that exploration: If a chemist can now craft self-reproducing molecular systems, what principles, if any, underlie the spontaneous formation of self-reproducing systems? If Ghadiri can build an autocatalytic peptide, or a collectively autocatalytic peptide reaction network, can such systems assemble themselves by chance? Is life prefigured in the laws of it all?

I broach a view, still theory, under which life, like Yeats's rough beast, doth slouch toward Bethlehem to be born—virgin birth of us all.

I wish to say that life is an expected, emergent property of complex chemical reaction networks. Under rather general conditions, as the diversity of molecular species in a reaction system increases, a phase transition is crossed beyond which the formation of collectively autocatalytic sets of molecules suddenly becomes almost inevitable. If so, we are birthed of molecular diversity, children of second-generation stars.

I begin with a toy example: buttons and threads. Consider 10,000 buttons on a hardwood floor and a spool of red thread. You pick a random pair of buttons, break off a length of red thread, and tie the two buttons together. Now, just repeat the process, picking successive random pairs of buttons, including ones that are already paired with another button, and tying them together with pieces of red thread. Every now and then, pause to lift a randomly chosen button off the floor and check how many other buttons you lift up with it as a single connected cluster of buttons.

Now here is the magic. At first, when there are 10,000 buttons and you have tied only a few pairs together, if you lift up a randomly chosen button, it will almost certainly be an isolated button, unconnected to any others by red threads. You might pick a button that is already a member of a pair or small cluster of a few connected buttons. But as you continue to tie more pairs of buttons together, the ratio of threads to buttons continues to increase. The size of the largest cluster of connected buttons will gradually increase. At some intermediate point, there are a modest number of modestly large connected clusters. But just at that point, addition of a few more threads will, by chance, connect buttons in several different modest-size clusters into a giant cluster (Figure 2.5).

In short, as the ratio of threads to buttons increases from zero, at first there are only small clusters of connected buttons. If one plots the size of the largest cluster (Figure 2.6) as the ratio of threads to buttons increases, the size of the largest cluster does not increase much at first, then quite rapidly increases to the giant cluster.

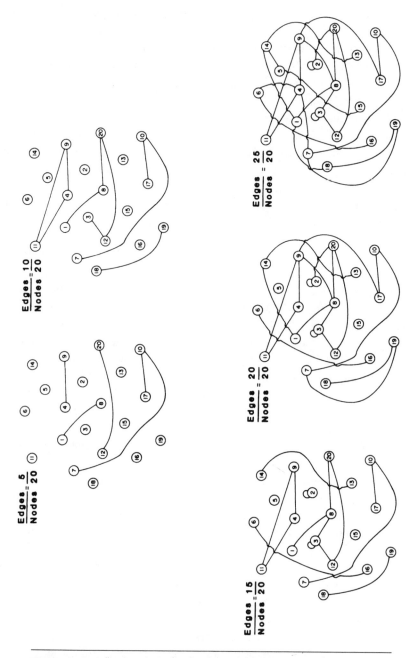

FIGURE 2.5 Crystallization of connected webs. Twenty buttons (nodes) are connected at random by an increasing number of threads (edges). For large numbers of buttons, as the ratio of threads to buttons increases past a threshold of 0.5, most points become connected in one giant component. As the ratio passes 1.00, closed pathways of all lengths begin to emerge.

With additional random pairing of buttons, gradually most of the remaining buttons are connected into the giant cluster.

Your task has just magically created a phase transition. If the number of buttons is very large, millions or more, then the steepness of the transition from tiny clusters to the giant cluster becomes an ever-more-sudden change at a critical value of the ratio of threads to buttons. The critical ratio is, in fact, 0.5, when the number of ends of threads, two per thread, is equal to the number of buttons.

The button-and-thread example, known more formally as a "random graph," was first studied by Hungarian mathematicians Erdos and Renyé in 1956. The central image you need to take away, where a graph is a set of points, or vertices, connected by a set of edges, is that a phase transition occurs. As the connectivity of a random graph increases, where the connectivity is just the ratio of edges to vertices —threads to buttons—a sudden jump to a richly connected network occurs at a critical ratio of edges to vertices.

It is my strong belief that a homologous phase transition to a giant connected component in chemical reaction systems underlies the emergence of collectively autocatalytic sets of molecules. Now that we have seen the button-and-thread case, the chemical case is easy to understand in outline.

The central vision is based on the concept of a "chemical reaction graph." One

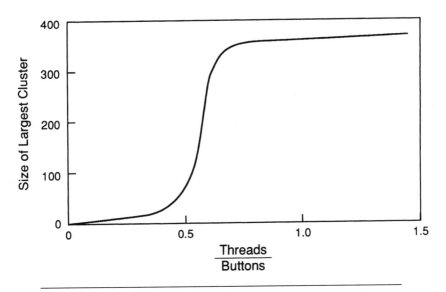

FIGURE 2.6 A phase transition. As the ratio of threads (edges) to buttons (nodes) in a random graph passes 0.5, the size of the largest connected cluster slowly increases until it reaches a phase transition and a giant component crystallizes. (For this experiment, the number of threads ranges from 0 to 600, while the number of buttons is held fixed at 400.)

need not be a chemist to know that in chemical reactions atoms or molecules enter as substrates, undergo transformations in which the substrate(s) are altered, and leave the reaction as product(s). A very simple categorization of chemical reactions is: one substrate–one product reaction; one substrate–two product reaction; two substrate–one product reaction; two substrate–two product reaction; higher-order reactions with more substrates or products.

A simple example of a one substrate–one product reaction is an isomerization reaction. The substrate and product differ by rotation of some of the atoms with respect to other atoms of a molecule around a single bond. A simple example of a one substrate–two product reaction is any cleavage reaction breaking a linear polymer into two fragments. For example, consider an RNA sequence, 3'CCC-CGGGG5', that is cleaved into two products, 3'CCC + CGGGG5'. A simple example of a two substrate–one product reaction is the inverse of the above example, ligating the two fragments, 3'CCC + CGGGG5', into the product, 3'CCC-CGGGG5'. And a simple example of a two substrate–two product reaction would take two RNA molecules, 3'CCCCC5' + '3GGGGG5', break both polymers at some internal point, then recombine the left end of the first polymer with the right end of the second, as well as the right end of the first with the left end of the second to yield, for example, 3'CCGG5' + 3'GGGCCC5'.

Note that in our hypothetical two substrate–two product reaction, two substrates with five nucleotides each entered the reaction, and two products, a shorter one with four nucleotides and a longer one with six nucleotides, left the reaction.

A reaction graph is only slightly different from a standard graph connecting buttons by threads or vertices by edges. Picture a set of, say, one thousand RNA sequences, considered as substrates and products of reactions, as dots, nodes, or vertices in a three-dimensional room (Figure 2.7a and 2.7b). Now connect any substrate(s) and product(s) of a reaction with a "hyperedge," consisting of lines leading from the substrate dots into a square box representing the reaction and lines leaving the square box and leading to the products. In order to distinguish substrates from products, it is convenient to place arrows on the lines directed from the box to the products. I emphasize that the arrows are for this convenience only and do not, as discussed just below, depict the direction of flow of chemical reactions. The set of all the substrate and product nodes, and all the hyperedges with boxes representing reactions, constitutes the chemical reaction graph (Figure 2.7a).

In principle, for any set of atoms and molecules, it is possible to construct a reaction graph showing the reactions that couple that set of molecules. However, it is important to emphasize that the reaction graph does not, as yet, tell us which way chemical reactions flow on the reaction graph. The direction of flow of matter on a chemical reaction graph depends upon the displacement from equilibrium across each hyperedge of the graph. The concept of chemical equilibrium is slightly subtler than the high school chemistry version, but that version will do. Consider a

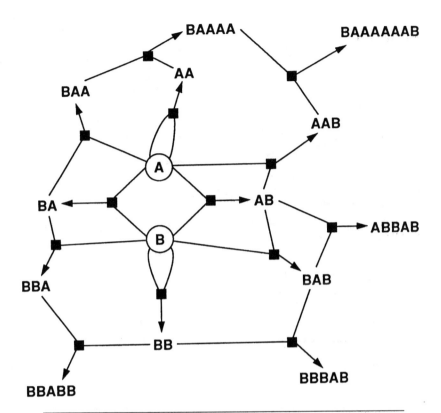

FIGURE 2.7a From buttons and threads to chemicals. In this hypothetical network of chemical reactions, called a reaction graph, smaller molecules (A and B) are combined to form larger molecules (AA, AB, etc.) which in turn are combined to form still larger molecules (BAB, BBA, BABB, etc.) Simultaneously, these longer molecules are broken down into simple substrates again. For each reaction, a line leads from the two substrates to a square denoting the reaction; an arrow leads from the reaction square to the product. (Since reactions are reversible, the use of arrows is meant to distinguish substrates from products only in one direction of the chemical flow.) Since the products of some reactions are substrates of further reactions, the result is a web of interlinked reactions.

one substrate–one product chemical reaction, AB, in an isolated, closed vessel. If A is present in high concentration and B is absent, A will convert to B at some rate that is proportional to the concentration of A. As the concentration of B increases, B will convert back to A at some rate that is proportional to the concentration of B. There is, therefore, some concentration of A and of B at which the rate that A converts to B is equal to the rate that B converts to A. That ratio of concentrations of A to B is chemical equilibrium. While individual A molecules continue to convert to B and B molecules convert to A, there is no net production of A or of B, beyond

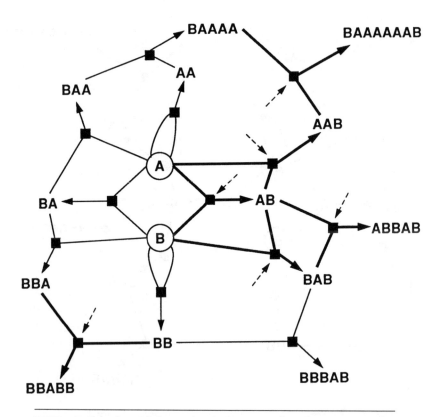

FIGURE 2.7b Molecules catalyzing reactions. In Figure 2.7a, all reactions were assumed to be spontaneous. What happens when we add catalysts to speed some of the reactions? Here the reaction squares indicated by dashed line arrows are catalyzed, and the heavy darker lines connect substrates and products whose reactions are catalyzed. The result is a pattern of heavy lines indicating the catalyzed subgraph of the reaction graph.

small fluctuations in A and B concentrations that tend to damp back toward the equilibrium ratio. Chemical reactions always tend, spontaneously, to proceed toward chemical equilibrium.

In an open thermodynamic system, our chemical reaction graph is open to the addition of matter and energy. For example, we might begin our reaction with some of the thousand RNA species present in the reaction vessel, we might persistently add certain specific RNA sequences to the vessel from the outside, and we might persistently remove certain sequences from the vessel. As a simple example, we might persistently add the four nucleotides of RNA, A, U, C, and G, to the vessel at a constant rate, and we might persistently remove all molecular species, A, U, C, G, and longer polymers, at a rate proportional to the concentration of each such species. A stirred chemostat is a convenient example of such a system. Here A, U, C, and G are added at a constant rate to a volume of buffer fluid and the buffer plus molecular mixture is removed from the vessel at a constant fluid volume rate, while new buffer is added to the mixture at the same volume rate. Among other experiments we might consider for such an open system would be to track the movement of the incoming A, U, C, and G into larger polymers, for example, by using radioactively tagged A, U, C, and G, then looking for radioactivity in the different larger polymers. Such an experiment would be tracking the flow of matter from the inputs to this open system, A, U, C and G, to many of the products of many of the reactions in the system. More generally, if we inject mass into the reaction graph at any point(s), we can study how it flows to other points in the reaction graph.

Now, on to the spontaneous emergence of collectively autocatalytic sets of molecules as another phase transition to a "giant component," in analogy to our buttons-and-threads picture. So far, we have pictured a chemical reaction graph in which all the reactions are uncatalyzed. So Figure 2.7a shows an uncatalyzed reaction graph. Each such reaction proceeds at its own pace and in a direction toward the equilibrium ratio of concentrations of substrate(s) to product(s) across that reaction hyperedge. But the pace of the spontaneous reaction might be very fast or very, very slow. Indeed, the pace of the spontaneous reaction depends upon the height of the energy barrier(s) separating the substrates from the products.

Physical chemists think of reactions in energy coordinates that reflect the total energy in a given molecule in a given configuration. That energy includes the energy involved in the bonds in the substrate(s), the distortion of those bonds from their at rest length and angles, entropic considerations, and so forth. In general, substrates and products are thought of as resting in energy wells (Figure 2.8), while the transition state molecular configuration, which occurs in passing from the sub-

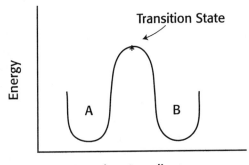

FIGURE 2.8 Two chemicals, A and B, in their respective potential energy wells, separated by a potential barrier. The transition state, *, occurs at the top of the energy barrier between the two energy wells.

strates to the products, consists of one or more molecules with distorted bonds and improbable spatial arrangements. Due to these distortions, the transition state occurs at a higher transition state energy compared to the energy of the substrates or products. The energy barrier for the reaction from substrate to product is the difference in height from the substrate energy to the transition state. The energy barrier of the back reaction from the product to the substrate is represented by the difference in height from the product energy to the transition state energy. (Indeed, if the heights are different from the substrate and product sides to the transition state, the "reaction rate" in the two directions is different, faster for the smaller energy difference to the transition state.)

Enzymes, and catalysts in general, speed up chemical reactions in the direction of equilibrium, but do not shift the chemical equilibrium itself. According to transition state theory, as mentioned previously, an enzyme works by binding to the transition state itself, thereby stabilizing the transition state. This stabilization lowers the energy of the transition state complex, which in turn decreases the energy barrier to the transition state from the substrate or from the product. In turn, this decrease in the energy barrier to the transition state increases the rate of reaction from the substrate toward the product and from the product toward the substrate, speeding up the reaction in both directions. So the catalyst speeds up the chemical reaction. And, because the speed up from substrate to product and from product to substrate is by the same multiplicative factor, the equilibrium ratio of substrate to product is not changed.

If some of the reactions in the uncatalyzed reaction graph in Figure 2.7a were catalyzed, we could show which reactions are catalyzed by coloring their corresponding hyperedges red, in analogy with the red threads linking buttons. Rather than coloring catalyzed reactions red, in Figure 2.7b the catalyzed reactions are shown by bold hyperedges. A reaction graph in which all catalyzed reactions are so depicted is a "catalyzed" reaction graph.

Now for the phase transition. Suppose we randomly pick hyperedges in a reaction graph and color them red, denoting that those reactions are catalyzed. At first, the red hyperedges are isolated. If one picks up a substrate at random, one can ask how many other substrates are connected to it directly or indirectly via catalyzed hyperedges. As more and more hyperedges are colored red, the number of red isolated hyperedges increases; then adjacent hyperedges are colored red, then linked sequences of hyperedges are colored red. In due course, a giant component of connected red hyperedges will arise. In short, initially most reactions are uncatalyzed, but as the ratio of catalyzed reactions to substrates and products increases, eventually a phase transition will occur, and a vast web of substrates and products are connected by the giant catalyzed hypergraph of reactions.

Once the giant catalyzed reaction web is present, matter added at points in the full reaction graph will tend to flow into and swirl around the rapid reactions in that giant web, and leak more slowly outward along the neighboring noncatalyzed reactions. If one thinks of the catalyzed reactions as major highways and the uncatalyzed reactions as rutted tracks leading from point to point, the matter in the chemical species added to the reaction graph at one or many points will tend to flow rapidly from substrates to products along the major highways, but trickle slowly along the rutted tracks.

What do we need to have a collectively autocatalytic set of molecules? Simple. We need a set of molecules having the property that, for each such molecular species, it either enters from the outside as "food" or has at least one final step in its synthesis catalyzed by one or more members of the collectively autocatalytic set. In our simplest example, Figure 2.3, A and B mutually catalyze as a last step in one another's formation from fragments of A and fragments of B, so the fragments of A and B are the "food" and A and B are the catalysts. But then, we need it to be true that the molecular species in our reaction graph are not only substrates and products, but also can serve as catalysts for the very reactions in the reaction graph.

Piece of cake. We already know that ribozymes are RNA sequences that catalyze the cleavage and ligation of RNA sequences. And we know that protein enzymes, such as the gut enzyme trypsin, cleave and ligate proteins—as when you digest hamburger. So RNA and proteins can be substrates and products and can also act to catalyze reactions involving RNA and proteins. More generally, RNA can act on proteins, proteins can act on RNA, both can catalyze reactions with other classes of molecules, and those other classes of molecules can catalyze reactions involving RNA and proteins.

Catalysis is ubiquitous.

If we knew which molecular species in our reaction graph catalyzed which reactions in our reaction graph, we could demark any such catalytic event both by coloring the corresponding hyperedge red and by drawing a blue arrow from the catalyst to the reaction "box" representing the reaction it catalyzed. If we could do that, then we could examine our catalyzed reaction graph, define our exogenous "food" molecular species, and see, by inspection, if the reaction graph contained one or more collectively autocatalytic sets.

In general, we do not, at present, know which "arbitrary" DNA, RNA, protein, or other molecules catalyze which reactions in an "arbitrary" chemical reaction graph. But initially we can do something that appears to be very useful. We can make alternative models, or hypotheses, about which molecules catalyze which reactions and then ask whether we would expect, on that basis, that a reaction graph would have collectively autocatalytic sets.

We need now only two more ideas. The first one notes that for linear polymers, like proteins and RNA, as the diversity of polymers in the system, and their mean lengths, increases, the number of reactions by which the molecules can be formed increases in diversity even faster.

Here is why. Consider RNA polymers of length six, such as 3'CCCCCC5' or 3'GCUUAA5'. Since there are 4 choices of nucleotides at each of six positions, the number of RNA sequences of length six is just four (the number of types of nucleotide building blocks) raised to the sixth power, hence 4096. But each such polymer can be built in five ways from its possible fragments. Thus, 3'GCUUAA5' might be constructed by ligating 3'G + 3'CUUAA5', or by ligating 3'GC + 3'UUAA5', and so forth. Since there are five internal bonds among the nucleotides, there are five ligation reactions that form each hexamer.

More generally, consider monomers, dimers, trimers, up to RNA hexamers, thus, A, U, C, G; AA, AU, AC... GGGGGG. A trimer, say, UAG, might be built up from smaller fragments in two ways, corresponding to its two internal phosophodiester bonds. But UAG might be cleaved from the 3' or 5' ends of tetramers, pentamers, or hexamers. For example, UAG might be cleaved from the 3' end of 3'UAGCCC5'.

Thus, in general, if we limit ourselves to cleavage and ligation reactions only, as the diversity of RNA sequences, monomers, dimers, and on up to Nmers increases, the number of reactions among those molecular species increases very much faster. Indeed, it is easy to show that the ratio of reactions to molecular species itself increases and is essentially proportional to the length of the longest polymers in the soup. That's intuitively rather obvious, as a hexamer can be made in five ways from smaller fragments, a heptamer can be made in six ways from smaller fragments, and so on.

Suppose our allowed list of chemical reactions includes two substrate–two product reactions, such as the recombination between two RNA sequences, 3'CCCCC5' + 3'GGGGG5' to yield 3'CCGG5' + 3'GGGCCC5'? For simplicity, suppose that each pair of different RNA molecular sequences could undergo only one such recombination reaction. Then the diversity of such reactions is equal to the number of distinct pairs of such RNA sequences. If there are N different sequences, then the number of pairs is N^2. So the diversity of reactions when recombination is included scales up as the square of the number of molecular species present. Hence the ratio of the diversity of reactions to the diversity of sequences is N^2 divided by N, hence N. But the ratio might be even larger, for our two initial polymers, 3'CCCCC5' and 3'GGGGG5', might have recombined at any of the four internal positions in either substrate, hence in sixteen ways, not just the particular single case noted above. In short, when two substrate–two product reactions are included, the ratio of reactions to molecular species explodes with the diversity of molecular species.

So we reach a general conclusion. Molecules are combinatorial objects. As the diversity of these objects increases, in general, the diversity of reactions among them increases much faster. In short, as the diversity of molecules in a reaction graph increases, the ratio of reactions, hence hyperedges, to molecules increases rapidly.

As the diversity of molecular species increases, the reaction graph becomes pregnant with the possibility of collective autocatalysis. There are soon many reactions per molecular species. A collectively autocatalytic set merely requires that at least one final reaction forming each member of the autocatalytic set be catalyzed by some member of that set.

What does it take to achieve a collectively autocatalytic set as the ratio of reactions to molecular species increases? If enough randomly chosen reactions are catalyzed by members of the reaction graph, one would expect, by chance, that a collectively autocatalytic set would eventually form.

Let's look at the question this way. We do not know which molecules catalyze which reactions. A zeroth-order simplest hypothesis would just allow a god of chance to assign at random which reactions each molecule catalyzes. In the simplest case, let's just assume that there is a fixed probability that any molecule can serve as a catalyst for any given reaction. For example, let that probability of catalysis, $pcat$, be 0.0000001. Then we build a model "catalyzed chemical reaction graph" by asking each molecular species in turn, with respect to each reaction in turn, "Do you catalyze this reaction or not?" We say "Yes," randomly, creating a probability of one in a million for a yes by drawing a random decimal uniformly between 0.0 and 1.0 on a computer and saying yes if it is less than 0.0000001. If the answer is positive, we draw a blue arrow from that lucky molecule to the lucky reaction it catalyzes, and we color the corresponding reaction hyperedge red.

So now it is easy to see what will happen. As the diversity of molecular species in our reaction graph increases, at some point the ratio of reactions to molecular species will be a million to one, while the probability any molecule catalyzes any reaction is one in a million. Hence, by chance, about one reaction per molecular species in the reaction graph will be catalyzed. The number of red hyperedges is now about equal to the number of molecular species.

Will a giant connected web of catalyzed reactions crystallize via our now-familiar phase transition from unconnected to connected at this point? Typically, yes! Typically, when there is about one catalyzed reaction per molecular species, a giant catalyzed web arises. And at that point, the system almost certainly contains one or more collectively autocatalytic sets of molecular species, for by chance it is now highly likely that the very giant web of catalyzed reactions leads from chosen input "food" molecules, via a sequence of catalyzed reactions, to the very set of molecules that catalyze the reactions in question.

As a very crude estimate, suppose the probability that a random RNA sequence catalyzes a randomly chosen reaction with two RNA substrates and two RNA products is one in a trillion, 10^{-12}. And allow that two RNA polymers length $L = 101$ can undergo recombination reactions in 100 x 100 different ways, then the diversity, N, of RNA sequences of length 101 to achieve autocatalysis by chance would be on the order of 10,000. The calculation is easy. The total diversity of recombination reactions is $(L - 1)(L - 1)(N)(N)$. Setting this as equal to a trillion, such that the probability that one reaction per molecular species finds a catalyst among the set, yields the result.

If this view is right, the emergence of autocatalytic sets is not hard, it is relatively easy. A way is needed to assemble varieties of, say, RNA or protein or other potential substrates and catalysts, hold them in proximity so they do not diffuse out of effective contact with one another, and let chance and number do their magic. Figure 2.9 shows a typical autocatalytic set.

If so, life is an expected emergent property of complex chemical reaction networks. But there are caveats: First, we need to know whether our crude calculation based on a simple pcat is robust. It appears to be. For a variety of hypotheses about the distribution of catalytic activities among sets of molecules, and a variety of hypotheses about the statistical structure of reaction graphs, autocatalytic sets tend to

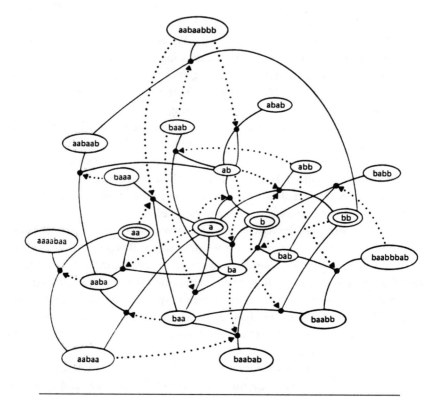

FIGURE 2.9 Atypical autocatalytic set, build up from a food set, concentric ellipses, consisting in the monomers, a and b, as well as the dimers aa and bb. Catalysis is shown by dotted lines leading from the catalyst to the reaction box, here reduced to a node, representing each of the ligation reactions.

emerge at a critical diversity. But caveats, more theory, and, most important at this stage, much more experimental work is needed.

Second, even if the above theory be true, we have not yet talked about the emergence of a metabolism that solves the thermodynamic problem of driving the rapid synthesis of molecular species above their equilibrium concentrations by linking such synthesis to the release of energy by the breakdown of other chemical species. Chemical reactions that release energy are called "exergonic." Reactions that require chemical energy are called "endergonic." Living cells link endergonic and exergonic reactions in order to build up high concentrations of many molecular species. I return to this central issue below, for the linking of exergonic and endergonic reactions appears to be essential to the definition of an "autonomous agent," that mysterious concatenation of matter, energy, information, and something more that we call life.

In short, I shall say, autocatalysis and molecular reproduction are necessary for life, but not yet sufficient. Life possesses deeper, still more mysterious properties than the autocatalysis we have explored in the present chapter.

I want to discuss two more issues before bringing the current chapter to a close: (1) the chemical adjacent possible; and (2) contemporary experimental approaches to test for the emergence of autocatalytic molecular systems. I didn't tell you the whole truth about chemical reaction graphs. They are much more interesting than Figures 2.7a and 2.7b.

Consider again our set of 1,000 organic molecules as substrates of reactions. It will do to consider 1,000 different RNA sequences, monomers, dimers, trimers, up to hexamers in length. Now consider all the possible cleavage, ligation, and recombination reactions among this initial "founder set" of organic molecules. In general, the products of that complete set of reactions will include molecular species that are not present in the founder set. For example, two hexamers might recombine to yield, as seen, a tetramer that might already exist in the founder set and an octamer that does not exist in the founder set. Consider the founder set as the "actual." Now consider the molecular species that are one reaction step away from the actual, but do not yet exist. Call this new set the chemically "adjacent possible."

Leap to the biosphere as a whole. There is some set of molecular species in our current biosphere. There is some further set of molecular species that does not exist in our biosphere, but is one reaction step away from the current actual in the chemically adjacent possible to our biosphere.

Four billion years ago, the chemical diversity of the biosphere was presumably very limited, with a few hundred organic molecular species. Today the biosphere swirls with trillions of organic molecular species. Thus, in fact, sunlight shining on our globe, plus some fussing around by lots of critters, has persistently exploded the molecular diversity of the biosphere into its chemically adjacent possible.

The adjacent possible will emerge as an important concept.

The species diversity of the biosphere has increased as well as the molecular diversity. The diversity of goods and services in our economy is huge compared to the diversity of goods achievable by Paleolithic humans 200,000 years ago. In short, the past four billion years has seen a persistent flow of the biosphere into its adjacent possible. Whence, why, and how all this burgeoning diversity? Our biosphere has and is plunging into its molecular, morphological, and behavioral adjacent possible. Moreover, as I will discuss further shortly, the universe has not had time to create all possible proteins of length 200 since the big bang. On the complexity scale of molecules, morphologies, species, Chevrolets, and legal systems, the universe is overwhelmingly nonrepeating, or "nonergodic." Indeed, I will be so bold below as to suggest that this nonequilibrium flow into a persistent adjacent possible may be the proper arrow of time, rather than the more familiar appeal to the second law of thermodynamics in closed thermodynamic systems. In addition, I will wonder whether there may be general laws governing any biosphere, part of a general biology, that govern how any biosphere persistently pierces and invades its adjacent possible. Finally, will we make collectively autocatalytic sets of molecules? Of course. Gunter von Kiedrowski has already achieved such a synthesis.

Can we test the hypothesis that as the diversity of molecular species increases the probability of collective autocatalysis increases sharply due to a phase transition in catalyzed reaction graphs from disconnected to connected? Almost certainly we will be able to in the near future. The bubbly field of molecular diversity, generating trillions of random DNA, RNA, proteins, and hundreds of thousands of small molecules in the search for drugs, will soon shift to the study of the collective properties of complex reaction networks with hundreds or thousands of such molecular species. Indeed, Andy Ellington, Reza Ghadiri, I, and others have proposed just such experiments to NASA. Whether we or others carry out the work, it cannot be long before we begin to explore experimentally the behavior of matter and energy linked in labyrinthine chemical reaction graphs. It cannot be long before we know how hard it really is to generate self-reproducing molecular systems.

If these views are right, it cannot be too hard to succeed. Life presumably did so without chemists on this wobbly ball of cooling rock some 4.3 billion years ago or more.

A general biology does indeed await us.

Chapter 3

AUTONOMOUS AGENTS

S OME WELLSPRING of creation, lithe in the scattered sunlight of an early planet, whispered something to the gods, who whispered back, and the mystery came alive. Agency was spawned. With it, the nature of the universe changed, for some new union of matter, energy, information, and something more could reach out and manipulate the world on its own behalf. Selfish? Yes. But how does matter, energy, information, and something else miraculous become selfish? From that miracle grew a biosphere—and, we must surmise, from that grow other biospheres, scattered seeds and gardens across the cosmos.

Pregnant in the birth of the universe was the birth of life. Agency may be coextensive with life. Life certainly burgeons nowhere without agency. We all act on our own behalf. In the Kantian form: What must something be such that it can act on its own behalf?

I will hazard again my tentative answer, baldly now, then return to it: An autonomous agent must be an autocatalytic system able to reproduce and able to perform one or more thermodynamic work cycles.

The thrashing *E. coli* swimming upstream in a glucose gradient is an autocatalytic system able to reproduce. In its swimming, it is carrying out one or more thermodynamic work cycles. Can I deduce this tentative definition from some more primary categories? I do not know how. At the beginning, I jumped—more realistically, I struggled—for weeks searching for a constellation of properties that seemed promising, in order to articulate something I sensed but could not yet say. It has taken four more years to understand more of what I still can only partially say in these investigations.

Definitional Jumps and Circles

What of definitional jumps in science?

So a pause to look at such jumps, to new clusters of concepts that, whole cloth, change how we carve up the world. Consider Isaac Newton and his famous $F = MA$, force is equal to mass times acceleration. It all seems sensible. If a log lies on an iced pond and I push it, it accelerates. If I push harder, it accelerates faster. If there are two logs, one on top of the other, and I push, the two accelerate less rapidly than if there were one log.

"I make no hypotheses," stated Newton. Yet Poincaré, two centuries later, argued that Newton's brilliant move was definitional. The three terms, F, M, and A—force, mass, and acceleration—argued Poincaré, admit no independent definitions. Acceleration is independently definable by the metric concepts of distance, motion, time, the rate of change of position with time—velocity, and the rate change of velocity with time—acceleration. But force and mass are, said Poincaré, joined at the hip, codefined, one in terms of the other. Mass is that which resists acceleration by force. Force is that which induces acceleration when applied to a mass.

The equation $F = MA$ is a definitional circle, according to Poincaré. Many physicists disagree with Poincaré. I tend to agree, but then I am not a physicist, so take care.

On the other hand, the great twentieth-century philosopher Ludwig Wittgenstein said something similar in his majestic *Philosophical Investigations*. Wittgenstein came to his revolutionary investigations painfully, ultimately rejecting his own earlier triumph in the *Tractatus*. Indeed, part of why I have so blatantly borrowed Wittgenstein's title, without my presumption to similar intellectual stature, is that there is a parallel between Wittgenstein's abandonment of the *Tractatus* and growing awareness of knowing as living a language game, and a central theme of this *Investigations*, that there may be a limit to the way Newton taught us to do science and a need to reformulate what we do when we and other agents get on with living a life. As we will see, in both cases, it appears that something profound is going on in the universe that is not finitely prestatable. Life and language games seem persistently open to radical innovations that cannot be deduced from previous categories and concepts. I will only be able to discuss these issues more fully after we have encountered autonomous agents and their unfolding mysteries.

In his early *Tractatus*, Wittgenstein had brought to conclusion the mandate of logical atomism from Russell. Logical atomism sought a firm epistemological foundation for all knowledge in terms of privileged "atomic statements" about "sense data." The idea, flowing from Hume to Berkeley to Kant then back to British empiricism, was the following: One might be mistaken in saying that a chair is in the room, but one could hardly be mistaken in reporting bits and pieces of one's own awareness. For example, "I am having the experience of a brown color"; "I am aware of the note A-flat." These statements were thought to record sense data and to be less susceptible to error than statements about the "external world" of chairs, rocks, and legal systems. Logical atomism sought to reconstruct statements about the external world from logical combinations of atomic statements about sense data. The *Tractatus* sought to bring that program to fruition... and nearly succeeded.

It was Wittgenstein himself who, twenty years later, junked the entire enterprise. *Philosophical Investigations* was his later-life revolution. His revolution has done much to destroy the concept of a privileged level of description and paved the way to an understanding that concepts at any level typically are formed in codefinitional circles. Newton was just ahead of his time.

You see, even in the case of the physical chair, the idea that we can reconstruct statements about chairs as equivalent to combinations of statements about sense data fails. The failure is deep and will have unfolding resonances with the mysteries of autonomous agents. The program to replace statements about physical objects with sets of statements about sense data requires "logical equivalence." Logical equivalence means that wherever a statement about a chair occurs at a "higher level" of description, it must be possible to specify a list of statements about sense data whose truth are jointly necessary and sufficient for the statement about the chair to be true. For example, the statement, "There is a Windsor rocker in the living room," must be implied by, "I seem to be seeing a brown surface with a given shape," "I feel a hard surface," and also, "When I push the top of the shape, I observe an oscillatory motion of the brown surface."

The trouble is, it appears to be impossible to finitely prespecify a set of statements about sense data whose truth is logically equivalent to statements about chairs. What if the observer were colorblind, or blind, or deaf, or not in the room but might enter the room wearing hobnailed boots, fall on the chair, and break it? Could we finitely specify the list, "If I were to enter the room, I would have the following sense data; and if one who is red-green colorblind were to enter the room he or she would have the following sense data; and if the rocker were to be on a soft rug, the sounds that would be heard by one who is A-flat tone deaf would.... To achieve this end, we would have to be able to finitely prespecify something about a set of statements concerning atomic sense data statements whose truth would be necessary and sufficient to the truth of a statement about the Windsor chair in the living room.

The problem, briefly, is that there appears to be no finitely prestatable set of conditions about sense data statements whose truth is logically equivalent to any statement about a real physical chair in a living room.

Wittgenstein invented the concept of a "language game," a codefined cluster of concepts that carve up the world in some new way. Consider, he said, legal language, and try translating it to ordinary statements about human agents without using legal concepts. So consider, "The jury found Henderson guilty of murder." We understand this statement but do so in the context of law, evidence, legal responsibility, trials, guilt and innocence, jury systems, bribing jury members, appeal processes, and so forth. Now try to translate the statement into a set of statements about ordinary human actions, "A group of twelve people were seated behind a wooden enclosure for several days. People talked about someone having died of poison. One day, the twelve people left the room and went to another room and talked about what had happened. Then the twelve people came back and one man stood up and uttered the words, 'We find Henderson guilty of murder.'"

There is no finitely prespecifiable list of statements about ordinary human actions whose truth would be necessary and sufficient for the truth of the statement that the jury found Henderson guilty of murder.

Another example concerns the United Kingdom going to war with Germany. There is no finitely prespecifiable set of ways this must happen. Thus, the queen, one day after tea, might betake herself to Parliament and cry, "We are at war with Germany!" Or the prime minister might in Parliament say, "Members, by the authority of my office I declare that we are at war with Germany!" Or ten thousand irate football fans from Leeds might shoulder picks, take to barges, funnel in fury up the Rhine, and attack football fans in Koblenz. Or . . . . In short, there is no prestatable set of necessary and sufficient conditions about the actions of individual humans for the United Kingdom to manage to go to war with Germany. (And may neither nation again find a way to make war on the other in the future.)

Wittgenstein's point is that one cannot, in general, reduce statements at a higher level to a finitely specified set of necessary and sufficient statements at a lower level. Instead, the concepts at the higher level are codefined. We understand "guilty of murder" within the legal language game and thus in terms of a codefining cluster of concepts concerning the features noted above—law, legal responsibility, evidence, trial, jury.

Useful new concepts arise in codefining clusters. It is fine that $F = MA$ is circular, Poincaré might say, for the codefining circle is at the center of a wider web of concepts, many of which have reference to the world; hence, the web of concepts touches, articulates, discriminates, and categorizes the world.

So too did Darwin jump to a new definitional circle with his concept of natural selection, heritable variation, and fitness. Many have worried that natural selection and fitness, which is defined as an increased expected number of offspring, are a definitional circle. Like $F = MA$, the definitional circle is virtuous.

So too I jump to the tentative definition, "An autonomous agent is a system able to reproduce itself and carry out one or more thermodynamic work cycles." Actually, at this stage my own tentative definition is not circular because "reproduce itself" and "work cycle" are definable independently. But as we delve further into the concept of an autonomous agent in succeeding chapters, cyclic definitions arise concerning "work," "propagating work," "constraints," "propagating organization," "task." I hope that the circle is virtuous and brings us toward a new understanding of "organization" itself.

In short, unpacking this definition will lead us into odd territory. Part of the oddness is the question of just what is the proper mathematical form to describe an autonomous agent. Is it a number, hence, a scalar? A list of numbers, hence, a vector? A tensor? I think not. An autonomous agent is a relational concept. In a later chapter I will suggest a spare mathematical form for an agent derived from category theory. This attempt is better than nothing, but I am not persuaded that effort is satisfactory, for the mathematical mappings of category theory are themselves finitely prestatable and a biosphere, I both fear and celebrate, is not. Somethings new are needed.

Autonomous Agents

Well, let's have at it. I begin with the cornerstone of thermodynamics, the Carnot cycle.

Sadi Carnot was a French engineer writing in the 1830s. The French had recently lost the Napoleonic Wars, Waterloo had come and gone, a certain famous statue to Wellington had been erected in London. Carnot, like others at the time, realized that part of the success achieved by the English had to do with early industrial economic power. The Brits had large numbers of steam pumps that were able to clear water from coal and iron mines, allowing miners to work under bad but relatively dry conditions and extract coal and iron more effectively than could their French counterparts. The coal and iron in turn were worked into the machinery of the Industrial Revolution and cannons and also, notably, more steam pumps that could keep the mines in working condition. I find it fascinating that the British system was an autocatalytic technology, iron and coal fueled the manufacture of pumps that abetted the mining of iron and coal.

Carnot set about understanding the fundamentals of the extraction of mechanical work from thermal energy sources. The result of his effort was an analysis of an idealized device to extract mechanical work from heat, the Carnot cycle. Carnot titled his work "An Investigation on the Motive Force of Heat."

A second point of fascination is that Carnot carried out his crucial analysis just as steam was becoming established as a source of mechanical work. It is almost certainly not a coincidence that the impulse to investigate the central nature of autonomous agents arises just as we are on the threshold of creating such molecular systems. Science, technology, and art tumble into the adjacent possible in roughly equal and yoked pace.

Figure 3.1 shows the essentials of Carnot's idealized machine. There are two heat reservoirs, one hotter than the other, $T_1 > T_2$. Between the two heat reservoirs is a cylinder with a piston inside. The space between the top of the piston and the head of the cylinder is filled with an idealized "working gas," which can be compressed and can expand. As a real, and certainly an idealized, gas is compressed, the gas becomes hotter; as the gas expands, it becomes cooler.

I will modify the Carnot machine in one central way, which does not alter but rather makes explicit an important feature of the actual working of the Carnot machine. I will attach a red handle to the cylinder, as shown in Figure 3.1. The Carnot cycle begins with the piston pressed up near the top of the cylinder, the working gas is compressed and hot, indeed, it is as hot as the high temperature T_1. You will operate the machine. You pull on the red handle, sliding the cylinder (frictionlessly) into contact with the high-temperature reservoir, T_1. You then let go of the red handle and allow the working gas in the cylinder to expand, thereby pushing the piston downward away from the head of the cylinder, in the first part of the power stroke of the Carnot engine.

As the power stroke initiates, the working gas expands and starts to cool. However, thanks to the fact that the cylinder is now in contact with the hot thermal

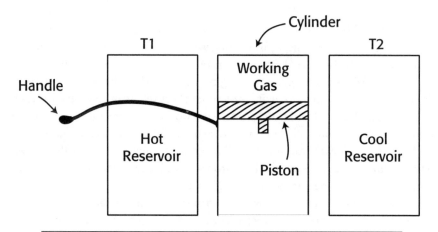

FIGURE 3.1   The essential machinery of the Carnot engine, comprising a hot reservoir at temperature T1 and a cool reservoir at temperature T2, T1 > T2. Between them is a cylinder and piston. Working gas is located in the space between the cylinder head and the piston. A red handle is attached to the cylinder, which you will use during the work cycle (shown in Figure 3.2) to move the cylinder between three locations, adjacent to the hot reservoir, a midposition between the two reservoirs, and adjacent to the cool reservoir, see text.

reservoir, heat flows into the cylinder from T_1 and maintains the working gas almost at the constant temperature T_1. Indeed, if the Carnot engine is operated slowly enough, the temperature remains constant. Such slow operation is said to be reversible. If the engine is operated more rapidly, hence irreversibly, the temperature is held nearly constant during this part of the power stroke, thus this section of the power stroke is called the "isothermal expansion portion" of the Carnot work cycle.

It is not only convenient but central to understanding a Carnot engine to plot the state of the system in a Cartesian coordinate system in which the x-axis corresponds to the volume of the working gas, while the y-axis corresponds to the pressure of the gas (Figure 3.2). The work cycle began with the piston near the top of the cylinder, the working gas hot and compressed. As the isothermal expansion

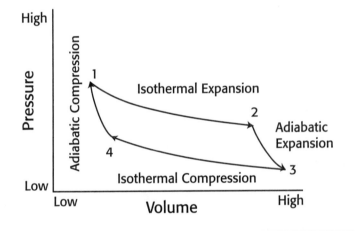

FIGURE 3.2 The Carnot work cycle in the Cartesian coordinate space of pressure and volume. The work cycle starts in state 1, with the cylinder adjacent to the hot reservoir, the piston pushed near the top of the cylinder, and the working gas hot and compressed. During the isothermal expansion, or power stroke stage, of the work cycle the working gas expands against the piston. Influx of heat from the hot reservoir keeps the working gas hot. The system flows from state 1 to state 2. During the adiabatic part of the power stroke, you have pushed the cylinder out of contact with the hot reservoir to the midposition. Further expansion of the working gas pushes the piston downward and completes the power stroke at state 3. To recompress and reheat the gas while extracting net work from the cycle, you push the cylinder into contact with the cool reservoir, then push on the piston from below, recompressing the gas. The heat generated by recompression is dumped into the cool reservoir during this isothermal part of the compression stroke, passing from state 3 to state 4. At state 4, you pull on the red handle to move the cylinder to the midposition between the reservoirs, then push up on the piston to further compress and heat the gas, returning the system to state 1. Total work performed is proportional to the area within the rhomboid figure connecting the four states.

phase of the power stroke happens, the pressure falls slightly, while the volume increases appreciably. The corresponding segment of the work cycle connects the starting position in Figure 3.2, position 1, to position 2 by a line segment representing the simultaneous values of volume and pressure during the isothermal expansion part of the power stroke.

You initiate the second part of the power stroke by pushing on the handle and moving the cylinder out of contact with the hot T_1 reservoir to a position between the two heat reservoirs, touching neither. You immediately let go of the handle. The working gas continues to expand, pushing downward on the piston and moving it away from the head of the cylinder. However, because the cylinder is out of contact with T_1 and the working gas is expanding, the working gas gets noticeably cooler, so pressure drops appreciably while volume increases slightly. This part of the power stroke is called "adiabatic expansion." In Figure 3.2, the adiabatic expansion step carries the system from step 2 to step 3, the end of the power stroke, a point where the pressure is at the lowest point, and the volume of the working gas is at the highest point of the work cycle.

Now in order to return the Carnot engine to its initial state, 1 (in Figure 3.2), such that the working gas can again expand and do mechanical work on the piston, work must be done on the engine to bring the piston back up to its position near the head of the cylinder and to recompress and reheat the working gas so that its temperature and pressure values, or state, correspond to state 1. If the work done on the Carnot engine simply retraced the exact pathway from step 3 to step 2 to step 1, at least as much work would have to be done on the Carnot engine as the engine released in the power stroke in going from 1 to 2 to 3. If so, no net work would be carried out by the Carnot engine on the external world.

Instead of retracing the power stroke pathway, the Carnot engine, and all heat engines, use a simple trick. You will do it. At step 3, the end of the power stroke, you push on the handle, pushing the cylinder into contact with the low-temperature reservoir, T_2. Indeed, you have arranged things such that at the end of the power stroke the working gas is itself at this lower temperature, T_2.

With the cylinder in contact with T_2, you now run around to the base of the cylinder, where a stout handle is attached to the piston and extends beyond the cylinder's base. You push on the handle, pushing the piston upward in the cylinder, thereby compressing the working gas. As you do this work on the piston, the compressing gas tends to heat up. But thanks to contact with the low-temperature reservoir, the heat generated by compression in the working gas diffuses into the cool T_2 reservoir, holding the working gas only slightly warmer than T_2. Thus volume decreases appreciably, while pressure increases slightly.

The point of this trick is that it requires less work to compress a gas that remains cool than one that heats up. Because the working gas is held at nearly a constant temperature T_2, this part of the compression stroke is called "isothermal compression," and it carries the system in its volume-pressure state space from position 3 to position 4 in Figure 3.2.

At the end of the isothermal compression stroke, you do your part again. You pull on the handle, pulling the cylinder out of contact with the cool T_2 reservoir to a position between T_2 and T_1, touching neither. Then you return to push on the handle connected to the piston, further compressing the working gas. Due to the compression of the working gas and the fact that it is not in contact with the cool T_1 reservoir, the working gas heats up, so the pressure increases appreciably while the volume decreases slightly, as the gas is compressed until the initial state of hot compressed gas, step 1, is achieved. The work cycle is now complete. The connected set of four lines in Figure 3.2, from step 1 to 2 to 3 to 4 to 1, depicts the sequence of pressure volume states of the working gas around the work cycle.

I have emphasized the role of you and the red handle in carrying us through the work cycle. In a real engine, of course, the role of the red handle is carried out by various gears, rods, escapements, and other mechanical devices that serve an essential role: The handle and you, or the gears, rods, and escapements, literally *organize* the flow of the recurrent process. I will return to this organization of the flow of process in a machine or an autonomous agent. The Carnot cycle is involved with the organized release of thermal energy in achieving recurrent mechanical work. The organization of work is essential—and will be central—to thinking about what occurs in an autonomous agent. Indeed, part of what we need is a way of characterizing the organization of real processes in the nonequilibrium world. I do not think we have, as yet, an adequate concept of organization.

The first question I want to raise about the Carnot cycle is this: Why is it a "cycle"? There is a clear sensible answer. The Carnot cycle operates in a cycle, as does a steam engine, gasoline engine, or electric motor, precisely because at the completion of one cycle the total system is returned to the same initial state as at the start of the cycle. Because of this, the organization of the process, achieved by the actual gears, rods, and escapements that coordinate the relative motion of the parts of the engine, has returned to the initial configuration from which, with no further ado, the system can again perform a work cycle. Thus in the case of the Carnot engine, the system performs mechanical work on the piston in a work cycle that, in net, transfers heat from the hot reservoir to the cold reservoir. At the end of the cycle, the Carnot engine returns to the initial state with the piston raised and filled with hot compressed gas at the temperature of the hot reservoir. The Carnot engine is all set as a total organization to receive another input of heat energy and perform another work cycle.

Suppose that there were no cycle? For example, a cannon fires a cannonball that hits a standing steel beam that is knocked over and hits a lever arm tossing a ball on its opposite end into the air which then falls to the ground and stops. To get the contraption to do it again would require that somehow the cannon be reloaded, the steel beam be reset on end, the lever arm to be reset with a ball on its other end. Where is all this organization of processes and events to come from? I will return to these issues in the next chapter, but notice for now that the cyclic organization of processes in the Carnot, the steam, the gas, and the electric engines achieve the requisite organization precisely because the system works as a cyclic process.

A second issue concerns a well-known feature of the Carnot engine that is very much worth comment here. If the sequence of steps is run in the reverse directions, so that the engine is run from state 1 to 4 to 3 to 2 to 1, the Carnot engine is not performing as a pump at all. It is, instead, performing as a refrigerator. Run in the reverse direction, the Carnot engine uses mechanical work to pump heat from the cool reservoir, T_2, to the hot reservoir, T_1, making T_2 cooler.

Now, heat does not spontaneously flow from cooler to hotter objects, making the cooler objects even cooler. So two points leap to attention here. First, the same device, the Carnot engine, can be a pump or be a refrigerator, depending upon the order of the operations. In fact, I am cheating a bit, for the organization of gears and escapements would differ in achieving the sequence of states 1, 2, 3, 4, 1 versus 1, 4, 3, 2, 1. But essentially the same machine can perform two very different functions, or tasks, pumping in one case, cooling in the other.

The third point, of course, concerns spontaneous processes and nonspontaneous processes. It took more than fifty years from Carnot's work to really begin to understand thermodynamics—the first law, conservation of energy; the famous and mysterious second law and its handmaiden, entropy; the third law concerning a zero temperature at which molecular motions stop—and for the invention of statistical mechanics to connect thermodynamics and Newtonian mechanics.

Some processes occur spontaneously, some conceivable processes do not. One of the most obvious cases is that if a hot gas is in contact with a cold gas, the two will in due course come to be the same temperature. Heat spontaneously "diffuses" from the hot to the cold object, cooling the former, warming the latter. In statistical mechanics we think of "hot" as corresponding to atoms moving rapidly, hence, with high kinetic energy. When these high-kinetic-energy atoms interact with slower-moving sets of atoms, the collisions transfer kinetic energy to the slower-moving atoms, speeding them up and slowing down the faster-moving atoms. Eventually, the atoms in the two sets come to have the same statistical distribution of motions, hence kinetic energy and hence temperature, where "temperature" is now understood to correspond to the average kinetic energy of the particles in the system.

The apparently mysterious second law of thermodynamics states that the entropy of a system is either constant or increases. The modern understanding of entropy can be stated roughly with the help of the concept of a $6N$-dimensional phase space.

Consider an isolated closed thermodynamic system, say, an ideal gas in a thermos bottle. Let there be N gas particles in the bottle. Now, each particle is in motion in real three-dimensional space; therefore, we can pick an arbitrary three-dimensional coordinate system, with length, width, and height (x, y, and z). We can note the current position of any particular particle in the bottle at any moment on each of the three positional coordinates. In addition to having a position, each particle may be in motion, it may have a velocity and an associated momentum in some direction in the bottle. Using Newton's vectorial composition of forces rules, we can decompose the motion of the real particle into its motions in the x direction, the y direction, and the z direction. The momentum in each of these directions is just the mass of the particle times its velocity in that direction. Newton's vector composition rule says that we can recover the motion of the initial particle by constructing the obvious parallelogram that adds the x, y, and z velocity or momentum vectors back together.

So for each particle we can represent its position and momentum in three spatial directions by 6 numbers. We have N particles in the bottle, so we can represent the current positions and momenta of all N particles at any instant in time by $6N$ numbers. Different combinations of positions and velocities now correspond to different sets of $6N$ numbers. And, as the N particles in the bottle collide and exchange momenta, bouncing off in new combinations of directions with new combinations of velocities according to Newton's three laws of motion, the $6N$ numbers representing the system at each moment change in time through some succession of $6N$ numbers. If we think of all the possible values of the positions and velocities of the N particles in the bottle, that set of possible values is the $6N$-dimensional phase space of our system. The system starts at some single combination of $6N$ numbers, hence a single state in the phase space. Over time, as positions and momenta change, the $6N$ numbers change and the system flows along a trajectory in phase space.

Now here is the heart of the second law: Some of the positions in the $6N$-dimensional phase space correspond to states in which all the particles are more or less uniformly spread throughout the bottle and are moving with more or less the same velocities. Other positions in the $6N$-dimensional phase space correspond to unusual situations, in which all the particles are located near the top of the bottle, are along the walls of the bottle, are moving in the same direction in the bottle, and so forth.

Consider mathematically breaking up the $6N$-dimensional phase space into a very large number of tiny $6N$-dimensional "cubes" that together add up to the entire $6N$-dimensional phase space. If the cubes are small enough, then each cube corresponds to a quite similar set of states. One cube might correspond to all the particles flowing downward in the bottle, another cube might correspond to a near uniform distribution of positions and momenta.

It is easy to see intuitively that there are many more combinations of positions and momenta that correspond to nearly uniform positions of the particles and their motions scattered in all possible directions than there are combinations of positions and momenta that correspond to all the particles being located in a specific small region of the bottle. In order to quantify this intuition, statistical mechanics counts the numbers of small $6N$-dimensional cubes corresponding to each such "macroscopic" state of the gas. Vastly many more small cubes correspond to the "macrostate" in which particles are nearly uniformly distributed and are moving in all the possible directions with velocities bunched around an average velocity—hence kinetic energy—than for any other macrostate, such as the macrostate corresponding to all the particles being located near the top of the bottle.

We are almost home free. We need one more premise—the famous "ergodic hypothesis." This hypothesis asserts that the trajectory of states leading from the initial state in a long stringlike "walk" will, over time, spend as much time in any small $6N$ cube as in any other cube. In short, the ergodic hypothesis asserts—indeed, assumes—that the system will wander around its phase space such that after a long time it will have spent as large a fraction of its time in any one tiny cube as any other cube.

But now we *are* home free. Since vastly many more small cubes correspond to the nearly uniform distribution of particles, moving in all possible directions but bunched around an average velocity, it follows from the ergodic hypothesis that the system will have spent most of its time in this "equilibrium" macrostate.

The second law, in its modern understanding, is simply the statement that an isolated thermodynamic system will tend to flow away from improbable macrostates—corresponding to very few of our tiny $6N$-dimensional cubes—and flow into and spend most of its time in the equilibrium macrostate for no better reason than that that macrostate corresponds to vastly many small $6N$ cubes in the entire $6N$-dimensional phase space. The increase of entropy in the second law is nothing but the tendency of systems to flow from less probable to more probable macrostates.

The physical concept of entropy of a macrostate is understood, since Ludwig Boltzmann in the last century, to be proportional to the logarithm of the number of small $6N$-dimensional cubes that correspond to that macrostate. The increase of entropy in spontaneous processes is then the tendency to flow from macrostates comprised of a small numbers of $6N$-dimensional cubes, or "microstates," to macrostates comprised of a very many microstates.

Our next step in thinking about autonomous agents requires us to consider again the concept of a "catalytic task space" and the character of autocatalytic sets in the context of catalytic task space. In the first chapter I described the basic framework, due to Alan Perelson and George Oster, of an abstract shape space, where each point would represent a molecular shape. Shape space has at least the three spatial dimensions of length, height, and width, but in addition it has properties reflecting the features of clusters of atoms that contribute to an effective molecular shape, that is, to those features that collectively might be "recognized" by another molecule that bound the shape in question. Such additional molecular features may include electric charge, dipole moment, hydrophobicity, or other features. At present, it is a reasonable guess that shape space is between five and seven dimensional. If true, this would not mean that there are five familiar features, three spatial and two others, that constitute the dimensions of shape space. Rather, some odd combination of several physical properties—partially charge, partially dipole moment, or partially hydrophobicity and partially dipole moment—might be the dimensions that matter.

In any case, we are to consider a bounded shape space with maximum and minimum values for each axis. A shape is a point in shape space. Thus, a molecular feature on a virus antigen, an epitope, is a point in shape space. An antibody might bind that epitope and a family of similar shapes filling a ball in shape space. As remarked above, very different molecules can have effectively the same shape, so endorphin and morphine both bind the same endorphin receptor. A finite number of balls will "cover" shape space, and the immune repertoire of perhaps a hundred million antibodies may well cover shape space.

Catalytic task space, you recall, simply applied the concept of shape space to catalysis. A point in catalytic task space now represents a catalytic task. A given chemical reaction constitutes a catalytic task. As in shape space, similar reactions constitute similar catalytic tasks. As in shape space, different reactions can constitute essentially the same catalytic task. An enzyme covers some ball in catalytic task space, comprising the set of reactions it can catalyze. And as noted before, according to transition state theory a catalytic task corresponds to a catalyst binding the distorted, hence high-energy, molecular configuration corresponding to the transition state of a reaction with high affinity and binding the substrate and product states with, in general, lower affinity.

In terms of catalytic task space, what is a collectively autocatalytic set? Consider a simple case. Two peptides, A and B, form a collectively autocatalytic set if A catalyzes the formation of B from two of B's fragments, while B catalyzes A from two of A's fragments. Then consider two balls in catalytic task space, the first ball, covered by A, constitutes the catalytic task in which two fragments of B are ligated to form B. The second ball, covered by B, constitutes the catalytic task in task space in which two fragments of A are ligated to form A.

The first feature of a collectively autocatalytic set is what I call "catalytic closure." Every reaction that must find a catalyst, does find a catalyst. The formation of A requires B, and the formation of B requires A. It is important to notice that this closure in catalytic task space is not "local"; there is no single reaction in this collectively autocatalytic set that by itself constitutes the closure in question. In a clear sense, the catalytic closure is a property of the whole system.

A second feature to notice is that A and B as catalysts do not by themselves constitute the closure in question; A and B might catalyze a variety of reactions. In particular, if B is presented with the two "proper" fragments of A, call them A' and A", then B will ligate A' and A" to form A. But if B were presented with other substrates, say Q and R, then B might catalyze a reaction transforming Q and R into two other molecular species, S and T. Similarly, A, as a catalyst, will ligate two proper fragments of B, B' and B" to form B. But A, if confronted with two other substrates, say, F and G, might catalyze their ligation to form a single third molecule, H.

While the set A, B, A', A", B', B" is collectively autocatalytic, forming more A and B from a substrate pool of A', A", B', and B", it is not the case that the set A, B, Q, R, F, and G is collectively autocatalytic, for the products of the reactions catalyzed by A and B, namely S, T, and H, are not themselves the catalysts A and B.

In short, the closure of catalytic tasks requires specification of the catalytic tasks themselves plus the specific substrates whose products, here A and B, constitute the very catalysts that carry out the catalytic tasks in question.

The closure of an autocatalytic set and set of catalytic tasks has a kind of dualism. From the point of view of the molecules involved, the specific catalytic tasks constitute the avenues of release of chemical energy by which the molecular system reproduces itself. The tasks coordinate the flow of atoms among the molecules whereby the set reforms itself. From the point of view of the tasks, the molecular species manage to carry out the tasks repeatedly, with no further molecular species being necessary to carry out the tasks. The molecules carry out the tasks, the tasks coordinate, or organize, the processes among the molecules.

The coordination afforded by the catalytic tasks that are jointly present and fulfilled is highlighted if we recall that, in general, two molecular species, say, A' and A", might undergo a variety of different reactions that form, in addition to A, perhaps E, L, M, P, and other molecular species. The specific catalytic task that carries A' and A" to A in the presence of a catalyst, B, speeds that specific reaction in comparison to the alternative reactions forming E, L, M, and P. Thus, the closure of the catalytic tasks, substrates and catalysts, A, B, A', A", B', B", achieves a coordination, or organization, of the flow of matter and energy into the autocatalytic system.

The organization achieved by the closure of catalytic tasks is similar to the organization achieved by the gears and escapements together with the rest of the idealized Carnot engine. The flow of process is marshaled into an organized whole. In the case of the autocatalytic set, the set reproduces itself. It also seems worth stressing that this closure in catalytic task space is a new concept with real physical meaning. It is a matter of objective fact whether or not a physical reaction system achieves catalytic closure; the hypothetical AB system above, and any free-living cell, achieves catalytic closure.

A final preliminary will bring us to our attempted definition of an autonomous agent. This preliminary is based on the distinction, noted above, between spontaneous and nonspontaneous chemical reactions. At equilibrium, the net rate of formation or destruction of each molecular species is zero, aside from small fluctuations that damp out. Thus, if two molecular species, X and Y, interconvert, the equilibrium is attained at that ratio of X and Y concentrations at which Y converts to X as fast as X converts to Y. If the reaction is displaced in one direction, say there is a higher X concentration than the equilibrium ratio, then the spontaneous, or exergonic, reaction proceeds in the direction toward equilibrium that reduces the excess of X concentration (Figure 3.3).

All spontaneous chemical reactions, if coupled to no other source of energy, are exergonic. On the other hand, if some other free energy source is coupled to the reaction, the reaction can be driven "beyond equilibrium" by using some of the energy source. Reactions that are driven beyond equilibrium by addition of free energy are called endergonic. Thus X might convert to Y, and this reaction might be coupled to another source of free energy, such that the steady state concentration of Y is much higher than the normal equilibrium X:Y ratio (Figure 3.3).

In the Carnot cycle, completion of the work cycle involved the cylinder piston system doing exergonic work on the external world during the power stroke, then the outside world doing work on the cylinder piston system when you pushed on

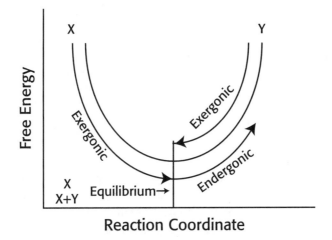

Reaction Coordinate

FIGURE 3.3 Free energy profile of a reaction in which x and y interconvert to an equilibrium ratio of x/(x+y). Free energy is minimized at equilibrium. If the system initially has an excess of x, or of y, with respect to equilibrium, the system flows exergonically down the free energy hill to the equilibrium position. If the system is coupled to an outside source of energy that drives the system beyond equilibrium—here to excess synthesis of y—the driving away from equilibrium is endergonic. Endergonic means that the reactions require the addition of free energy from some further source. Spontaneous processes are, overall, exergonic—they give off free energy.

the piston to recompress the working gas. The Carnot cycle links mechanical and thermal energy sources into a work cycle. A chemical reaction network with a work cycle will have to link spontaneous, exergonic and nonspontaneous, endergonic reactions into the chemical analogue of a work cycle. Like the cyclic Carnot engine, the chemical analogue will have to work in a cycle of states, like the 1, 2, 3, 4, 1 cycle of the Carnot cycle. Further, in order for the cycle to operate at a finite rate, hence irreversibly, the autonomous agent must be an open thermodynamic system driven by outside sources of matter or energy—hence "food"—and the continual driving of the system by such "food" holds the system away from equilibrium.

In this light, think again of the Ghadiri autocatalytic system, the 32-amino-acid sequence A that ligates two fragments A', a 15-amino-acid fragment, and A", a 17-amino-acid fragment, into A. This reaction is purely exergonic. The reaction proceeds from the substrate fragments A' and A" to form the product molecule A and approaches the equilibrium ratio of substrates to product. Ghadiri's autocatalytic system is wonderful, but merely exergonic. It does not achieve a work cycle. In general, autocatalytic and collectively autocatalytic systems can be purely exergonic. In any such case, no work cycle is achieved.

Now we can return to my jumped-to definition: An autonomous agent is a reproducing system that carries out at least one thermodynamic work cycle. That bacterium, sculling up the glucose gradient, flagellum flailing in work cycles, is busy as hell doing "it," reproducing and carrying out one or more work cycles. So too are all free-living cells and organisms. We do, in blunt fact, link spontaneous and nonspontaneous processes in richly webbed pathways of interaction that achieve reproduction and the persistent work cycles by which we act on the world. Beavers do build dams; yet beavers are "just" physical systems.

But Reza Ghadiri's example of an autocatalytic peptide doesn't make the grade, nor does Gunter von Kiedrowski's autocatalytic hexamer DNA or collectively autocatalytic set of two DNA hexamers. All these systems are merely exergonic. No work cycle is performed.

Now that we have stated our proposed definition of an autonomous agent, it is not too hard to imagine a chemical realization. In Figure 3.4 I show a hypothetical molecular autonomous agent. Given visualization of a first case, I expect that we will be constructing molecular autonomous agents within a few years.

Figure 3.4, our first example of a candidate molecular autonomous agent, is "constructed" to link with two further molecular systems, the exergonic autocatalytic system developed by Gunter von Kiedrowski based on ligation of two DNA trimers by their complementary hexamer. Here, the hexamer is simplified to 3'CCCGGG5', and the two complementary trimers are 5'GGG3' + 5'CCC3'. Left to its own devices, this reaction is exergonic and, in the presence of excess trimers compared to the equilibrium ratio of hexamer to trimers, will flow exergonically

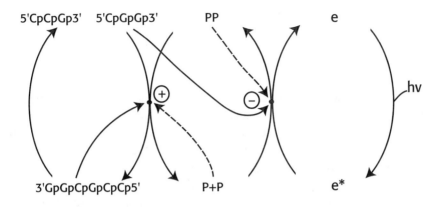

FIGURE 3.4 A first hypothetical molecular autonomous agent. The entire system is an open nonequilibrium thermodynamic system driven by two sustained sources of food. The sources of food are two single stranded DNA trimers and a photon source, *hv*. The two single stranded DNA trimers undergo a ligation reaction to form a second copy of the hexamer, thereby replicating the hexamer. There are two reaction pathways of synthesis of the hexamer and its breakdown back to the trimers. Each pathway is reversible, but due to displacement from equilibrium by the food, tends to flow in a forward direction toward synthesis of the hexamer. The first forward pathway is coupled to the breakdown of pyrophosophate, PP, to two monophosphates, P + P. The exergonic drop in free energy from the breakdown of PP drives excess *endergonic* synthesis, hence excess replication, of the DNA hexamer, compared to the equilibrium otherwise achievable. There is a reverse pathway by which the hexamer breaks down to the trimers that does not couple to PP. However, in the absence of an enzyme to catalyze this return reaction, the reaction is very slow, and the excess replicated DNA hexamer can persist in the system for a long time. PP, in turn, is reformed from religation of P + P by an *endergonic* reaction that is driven by coupling to a photon-excited electron, e*, which gives up free energy in an exergonic reaction to return to a ground state, while adding that energy to the endergonic synthesis of PP from P + P. The electron at ground state, e, is reexcited endergonically by absorption of another photon, *hv*. The hexamer catalyzes the reaction leading to its own excess synthesis. One of the trimers catalyzes the reaction leading to PP resynthesis. The analogue of gears and escapements in a Carnot engine is played by allosteric enhancers and inhibitors that gate the different reactions so they occur reciprocally. Thus, monophosophate, P, activates the hexamer enzyme. Pyrophosophate, PP, inhibits the trimer enzyme. Overall, the autonomous agent couples reproduction and a work cycle.

toward equilibrium by synthesizing the hexamer. Because the hexamer is itself the catalyst for the reaction, the synthesis of hexamer is autocatalytic.

The first additional system consists in pyrophosphate, PP, a high-energy dimer of monophosphate that breaks down to form two monophosphates, P + P. Like any reaction, the reaction converting PP to P + P has an equilibrium, hence an equilibrium ratio of PP and P. In the presence of excess PP compared to equilibrium, the reaction flows toward equilibrium by the spontaneous cleavage of PP to yield P + P.

My purpose in invoking the exergonic conversion of PP to P + P is to utilize the loss of free energy in this exergonic reaction to drive the DNA trimer-hexamer reaction beyond its own equilibrium, leading thereby to an *excess synthesis* of the 3'CCCGGG5' hexamer when compared to its equilibrium concentration. Thus, the excess synthesis of the hexamer, which would not occur spontaneously, is driven *endergonically* by being *coupled* to the exergonic breakdown of PP to P + P (Figure 3.4). In short, the exergonic breakdown of PP to P + P supplies the free energy to drive the excess buildup of 3'CCCGGG5' concentration beyond its own equilibrium with respect to its trimer substrates, 5'GGG3' and 5'CCC3'.

The excess synthesis of 3'CCCGGG5' constitutes excess reproduction of the hexamer autocatalytic reaction product beyond that which would occur without the coupling to the additional PP free energy source. Thus, the system is reproducing "better" with the coupling to PP than without the coupling.

Another point to note is that the coupling of the breakdown of PP to P + P with the excess synthesis of the DNA hexamer compared to the equilibrium concentration of the DNA hexamer means that energy is stored within the system. This is true because the excess concentration of the hexamer DNA, compared to its equilibrium, could in principle be released by degradation of the hexamer to the two trimer substrates, releasing that stored free energy as this reaction couple flowed toward its own equilibrium ratio of hexamer and trimers. Thus, the coupling to the PP to P + P reaction means that the autonomous agent stores energy internally. Later in evolution, such internally stored energy can be used to drive reactions that correct errors, as in DNA repair in contemporary cells. I am glad to thank Phil Anderson and, indirectly, John Hopfield for this point.

Once the pyrophosphate, PP, is cleaved to form P + P, as this reaction flows toward its own equilibrium ratio of PP to P, that free energy is used up. In order to have a renewed internal supply of the free energy needed to synthesize excess hexamer, it is convenient to resynthesize pyrophosphate from the two monophosphates, P + P. I'll return below to the meaning of "convenient," for in a general sense, the convenience reflects the organization of processes that sustains an agent, and that organization is not convenient, it is essential.

Resynthesis of PP from P + P requires the addition of free energy. This is true because we used the exergonic breakdown of PP to P + P to drive the excess synthesis of 3'CCCGGG5'. Now we need to add energy to resynthesize PP from P + P. To do so, I invoke an additional source of free energy in the form of an electron, e, which absorbs a photon, $h\nu$; is driven endergonically to an excited state, e^*, and falls back exergonically to its low-energy state, e, in a reaction that is coupled to the synthesis of PP from P + P.

The point of this third reaction-couple is clear: PP is resynthesized from P + P so that PP can continue to drive the excess synthesis of the DNA hexamer, 3'CCCGGG5'. Overall, the total system of linked reactions is exergonic—there is an overall loss of free energy that is ultimately supplied by the incoming photon, hv, plus the 2 substrates, 5'GGG3', and 5'CCC3'. Thus, we are not cheating the second law.

Let's return to the Carnot cycle, where I had you pushing and pulling on the handle and on the piston itself during the work cycle. We noted that in a real engine you would not be busy pushing and pulling. Your role in organizing the processes would be taken by gears, escapements, rods, connectors, bearings, and other bits of machinery.

I now invoke the analogue of the gears, rods, and connectors in the form of hypothetical molecular couplings that control the reactions I have already invoked. Specifically, I will assume that the hexamer, 3'CCCGGG5', is the catalyst that *couples* ligation of the two trimers, 5'GGG3' + 5'CCC3', to the exergonic breakdown of PP to P + P. My second assumption is that monophosphate, P, binds to the hexamer and facilitates the reaction. Thus, I am assuming that P is an allosteric enhancer of the reaction. "Allosteric" means that P binds to a site on the enzyme, here the hexamer, other than the hexamer's own binding site for the substrates. Allosteric enhancers and inhibitors are common in biological systems.

Here, P might bind to the sugar-phosphate backbone of the DNA hexamer. This coupling implies that as PP breaks down to form P + P, the monophosphate, P, will feed back to further activate the hexamer enzyme, making the catalysis of hexamer formation even more rapid. Just such a positive feedback of a reaction product on the enzyme forming it occurs in the famous glycolytic pathway that is the core of metabolism in your cells. In fact, under appropriate experimental conditions, this positive feedback coupling can cause the glycolytic pathway to undergo sustained temporal oscillations in the concentrations of the glycolytic metabolites.

Finally, I will invoke a few more couplings. I assume that one of the trimers, 5'CCC3', is the catalyst that couples the exergonic loss of free energy from the activated electron, e^* to e, with the resynthesis of PP from P + P. And I invoke an allosteric inhibition of this catalysis by PP itself. Thus, when PP is in high concentration, it tends to inhibit its own resynthesis. But when PP concentration falls, the inhibition on PP resynthesis is removed, and PP is resynthesized. The whole molecular contraption, our first hypothetical autonomous agent, is shown in Figure 3.4.

One of the first things to note about our hypothetical autonomous agent is that it constitutes a previously unstudied class of chemical reaction networks. The behavior of exergonic autocatalytic and cross-catalytic systems is beginning to be studied. The behavior of linked exergonic and endergonic reaction networks is the very stuff of intermediate metabolism and energy's biochemical transduction, studied for years by biochemists. But, to date, no one has begun to study linked reaction networks in which autocatalysis is coupled to linked exergonic and energonic reactions. So we are entering an entirely new domain.

Thus, our molecular autonomous agent constitutes a system with two essential features of living systems, self-reproduction and metabolism. However, my insistence that an autonomous agent carries out a work cycle refines the generally understood concept of a metabolism to include the requirement that the metabolism carries out a work cycle.

The second feature to note is that our autonomous agent is, necessarily, a nonequilibrium system. Free energy, here in the form of the photon, hv, and the trimer substrates is taken in and used to drive the linked synthesis of PP and excess DNA hexamer. There is no agency at equilibrium. The excess synthesis of DNA hexamer constitutes excess replication of the hexamer by virtue of the coupling of the trimer-hexamer synthesis to the PP $\leftrightarrow$ P + P cycle of reactions, which, as noted next, constitute a "chemical engine."

The third feature to note is the work cycle performed by the agent. The simplest way of seeing the work cycle here is in the behavior of the PP $\leftrightarrow$ P + P reaction. In the Carnot cycle, the working gas cycles from compressed and hot to less compressed and cool, back to compressed and hot. In our hypothetical autonomous agent, there is a macroscopic cycle of matter from PP to P + P via the reaction-forming DNA hexamer and back around to PP via the reaction with the high-energy electron. The macroscopic cycling of matter around this cycle is the engine at work. (I am grateful to Peter Wills for this clarification of the concept of a chemical motor.) In addition, depending upon the details of the kinetic constants, our autonomous agent may literally show an oscillatory concentration cycle in which PP concentration begins high and falls as P + P is formed, then the high PP concentration is reformed by use of the photon-energized exergonic $e^* \leftrightarrow e$ reaction.

Thus, the PP $\leftrightarrow$ P + P reaction embedded in the autonomous agent constitutes a chemical engine in which there is a macroscopic net flux of matter around the PP $\leftrightarrow$ P + P cycle, which is operating displaced from equilibrium as it is driven by addition of energy from the photon, hv, and addition of the two DNA trimers, and as energy is drained off to drive excess synthesis of the DNA hexamer.

The fourth thing to note about the autonomous agent is that, like the Carnot engine, the steam engine, the gas engine, and the electric engine, the autonomous agent works in a cycle. At the end of the cycle the system is poised to cycle again. A repeating organization of process is achieved. And next, just as the Carnot engine run backward is a refrigerator and not a pump, if the reactions of the autonomous agent were run backward the PP $\leftrightarrow$ P + P engine would run in the reverse direction. This is because all reaction couples would be displaced from equilibrium the opposite way and the analogue of throwing the gears in reverse, namely reversing in sign the positive and negative activator and inhibitor couplings to the two proper enzymes, would convert the excess energy stored in the above equilibrium concentration of hexamer into production of the two trimers and the resynthesis of PP from P + P. Were the release of the photon, hv, a readily reversible step, the excess of PP would drive emission of a photon by the excited electron, thus returning the electron to the initial unexcited state.

In short, if the autonomous agent is run backward, the autonomous agent melts down into its foodstuff. Run backward, the system is not an autonomous agent, for it does not reproduce itself and perform a work cycle. Run backward, the system is a flashlight!

Does the autonomous agent work? The answer is yes. My colleagues Andrew Daley, Andrew Girvin, Peter Wills, and Daniel Yamins and I have simulated the system of differential equations that correspond to the dynamics of this autonomous-agent molecular reaction network. The differential equations represent the way the concentration of each chemical species in the autonomous agent changes over time as a function of its own and other chemical concentrations. In general in such mathematical models, a number of unchanging constants representing kinetic constants and other parameters enter into the differential equations. In the present case, the differential equation system has thirteen such parameters.

The model autonomous agent system is displaced from equilibrium by the persistent addition of the two DNA trimers, 5'GGG3' and 5'CCC3', the removal of the DNA hexamer, and the persistent shining of photon, hv, from the outside. The chemical reaction network occurs under "chemostat" conditions. This means that all molecular constituents of the system are treated mathematically as if they were in a real well-stirred container to which the trimers and photons are added at a constant rate. In addition, the hexamer molecular components are removed from the system at an adjustable rate that holds their internal concentrations constant whatever the rate of reproduction of hexamer may be.

The autonomous agent system reproduces more efficiently with the couplings of the DNA trimer-hexamer system to the PP and electron cycles than in the purely exergonic case in which the DNA trimer-hexamer system operates alone. We measured efficiency thermodynamically as the conversion of available free energy coming into the system from the photon source into the excess hexamer with respect to the undriven steady-state rate of reaction concentration of the hexamer.

Figure 3.5 shows the results of our simulations. In these simulations of the chemical reaction network, there are, as noted, thirteen kinetic constants. We carried out computer selection experiments not only comparing the autonomous agent to a nude exergonic DNA trimer-hexamer system, but also computationally mutating the kinetic constants by small amounts and computationally evolving autonomous agents to reproduce with higher thermodynamic efficiency.

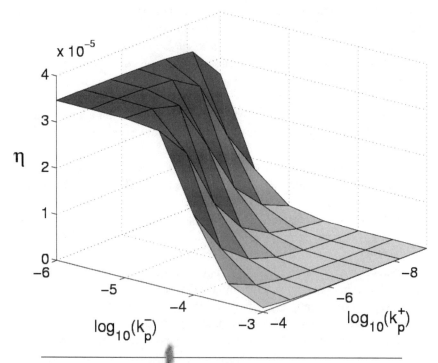

FIGURE 3.5 The autonomous agent outreplicates a simple DNA hexamer-trimer ligation system. Figure shows part of the fitness landscape in the thirteen-dimensional parameter space. Here fitness, measured as the efficiency of conversion of DNA trimer food sources into DNA hexamer, hence excess hexamer replication, is plotted on the vertical axis with respect to two of the thirteen kinetic parameters of the model molecular autonomous agent.

Our results demonstrate first and most important that autonomous agents operating displaced from equilibrium and utilizing a work cycle can be more efficient at using the available free energy coming into the total system in reproducing hexamer DNA than in the absence of the coupling of the trimer-hexamer DNA system to the PP and electron-photon work cycle system. Thus, the autonomous agent as a whole, including its work cycle, reproduces DNA hexamer more rapidly than would the trimer-hexamer exergonic system alone. In short, and also important, being an autonomous agent coupling an autocatalytic system with a work cycle is of selective advantage compared to being a merely exergonic autocatalytic system.

Second, just as in the glycolytic positive-feedback case, our autonomous agent model, for appropriate values of the kinetic constants, can undergo sustained temporal oscillations of PP and other concentrations. The oscillation of PP from high concentration to low concentration then back to high concentration during the work cycle is analogous to the expansion and recompression oscillation of the working gas in the Carnot engine's work cycle.

Third, a mountainous fitness landscape exists in the mathematical parameter space of the thirteen kinetic constants, in which some values of the kinetic constants lead to higher efficiency of reproduction than others. Darwin's natural selection could, in principle, operate if there were heritable variation in the kinetic constants.

The main conclusion to draw from our simulation is that autonomous agents coupling one or more autocatalytic and work cycles are a perfectly reasonable, if novel, form of nonequilibrium, open chemical reaction network. There is no hocus pocus here. In the near future we will almost certainly construct such autonomous-agent molecular reaction networks and study their dynamics and evolutionary behavior. A general biology is, in fact, around the corner.

The hypothetical molecular autonomous agent that we have considered has been discussed, for simplicity, as if the problem of retaining the reactants in a confined region of space could be ignored. In fact, this assumption is an idealization. Were our autonomous agent in a dilute solution, the rates of reaction would be very slow. Actual creation of a functioning molecular autonomous agent will require that the reacting molecular species be confined to a small volume or a surface or in some other fashion.

Candidates for isolation to small volumes include micelles and liposomes. Both macromolecular aggregated structures are comprised of "amphipathic molecules," that is, molecules with hydrophobic and hydrophilic regions such as lipids. Micelles are single-layered structures which, in an aqueous medium have hydrophilic regions directed outward, but are able to enclose an aqueous core in which the other molecular species of an autonomous agent might reside. In an aqueous medium, liposomes form double-layered membranes, homologous to cell membranes, with hydrophilic heads in the aqueous medium and hydrophobic tails abutted. In an aqueous medium, both micelles and liposomes can form and even reproduce by budding. A full-fledged molecular autonomous agent would have to synthesize the lipid or similar molecular constituents of its bounding surface and coordinate budding with dispersion of autocatalytic and work cycle partners to daughter vesicles.

An alternative to isolation of the autocatalytic and work cycle molecular species within a bounding volume is the confinement of such reacting species to a surface. Such confinement has the further advantage of altering diffusive search by reactants from a three- to a two-dimensional search process. The latter can shift the corresponding chemical equilibrium toward synthesis of larger polymers from their smaller substates. Here one can imagine confinement of reactants and products to clay surfaces or confinement of complex organic reactants and products to the surfaces of the abundant dust particles in giant molecular clouds in galaxies.

I will have much more to say in subsequent chapters about the properties of molecular autonomous agents. In particular, in order to understand agents we will have to carry out a critique of the physicist's concept of "work," as in a work cycle, for the best understanding of "work" appears to be that work is the constrained release of energy. Yet the very constraints on the release of energy that are essential to the doing of work themselves constitute the analogues of the gears, rods, connectors, and escapements of an ordinary machine. Most important, it typically takes work itself to construct the constraints on the release of energy that then constitutes work. In our first example of an autonomous agent, Figure 3.4, these constraints are present in the invoked couplings of catalysts and allosteric effectors to the reactions of which the autonomous agent is comprised. I have a hunch—a deep hunch verging on conviction—that the coherent organization of the construction of sets of constraints on the release of energy which constitutes the work by which agents build further constraints on the release of energy that in due course literally build a second copy of the agent itself, is a new concept, the proper formulation of which will be a proper concept of "organization."

In the meantime, if I am right, what did Schrödinger miss? He was right about his microcode—the microcode will reemerge as a subset of the constraints on the release of energy by which an autonomous agent builds a rough copy of itself. Namely, the microcode is the very structure of DNA, which serves as constraints on the enzymes that then transcribe and translate the code. But Schrödinger missed stating the requirement for an agent to be nonequilibrium. On the other hand, displacement from equilibrium is a necessary condition for a microcode to do anything at all. So perhaps displacement from equilibrium was implicit in his theme. More important, I think, is that he missed the concept that an agent is a union of an autocatalytic system that does one or more work cycles. This union is a new kind of dynamical system.

Now that we have seen an autonomous agent, I find myself wondering whether autonomous agents may constitute a proper definition of life itself. I make no attempt to defend my own strong intuition that the answer is yes. I suspect that the concept of an autonomous agent as an autocatalytic system carrying out one or more work cycles defines life. If so, here is the center, the elusive core of life, that examination of the molecular chunks of cells does not reveal. Most of the remainder of this book is devoted to examining the unexpected unfoldings of this tentative definition of autonomous agents and, perhaps, life. But I certainly will not insist upon my intuition. It suffices at this stage to note that all free-living systems we know—single-cell bacteria, single-cell eukaryotic cells, and multicelled organisms —fulfill my definition of autonomous agent.

If Figure 3.4 shows us a first case of a molecular autonomous agent, how broad a family of systems does the concept of an autonomous agent embrace? I confess I do not know. Clearly, there is nothing in the concept of a reproducing system that carries out at least one thermodynamic work cycle that limits such a system to DNA, RNA, and proteins. As we have seen, Julius Rebek has already created self-reproducing organic molecules well outside the familiar classes of biopolymers. If no such reproducing molecular system yet enfolds a thermodynamic work cycle, that is not to say that we shall long be stalled in creating such systems. It seems plausible that wide classes of chemical reaction networks can fulfill the criteria I have traced above. But must autonomous agents be "molecular" in the familiar sense? Could mutually gravitating systems such as galaxies fulfill the criteria? What of systems made largely of photons, self-reproducing spectra in a lasing cavity fed by a gain medium? What of geomorphology? I do not know. Perhaps it suffices at this stage to have begun an enquiry, an investigation, rather than to have completed it.

Natural Games

I turn, in the final section of this chapter, to yet another puzzle concerning what I call a natural game. A natural game is a way of making a living in an environment. That is, autonomous agents are able to act on their own behalf and regularly do so in order to make a living in an environment. The bacterium swimming upstream in a glucose gradient is making a living in its environment. So, in fact, are all free-living entities in the biosphere.

Well, it seems straightforward enough; we all know more or less what it is to make a living. For example, I am currently writing *Investigations* as part of my own hopefully not-too-solipsistic efforts to make my own living as a scientist.

But natural games are not quite so straightforward. I begin by mentioning again the rather surprising no-free-lunch theorem proved by Bill Macready and David Wolpert as postdoctoral fellows at the Santa Fe Institute a few years ago. Recall that Bill and David were wondering whether there might be some search algorithm for adapting on a fitness landscape that was inherently better than all other algorithms. For example, John Holland, another Santa Fe Institute colleague, is justly well-known for inventing his "genetic algorithm" to optimize hard computational problems. The genetic algorithm, which has been rather widely used in academic and industrial settings, is based on analogy with biological adaptation driven by mutation, recombination, and selection.

In effect, Bill and David were wondering whether biological systems in this biosphere happen to have stumbled on the best possible optimization procedure. Importantly, the answer appears to be no. Macready and Wolpert considered a mathematically well-formulated set of "all possible fitness landscapes" and showed that, averaged over all landscapes, no search algorithm outperforms any other algorithm. No free lunch.

In short, given an arbitrary fitness landscape, only some search algorithms do well on that landscape. The search procedure must be tuned to the fitness landscapes being searched if the search procedure is to be more effective than average among search procedures.

But this poses the important problem raised in chapter 1. Most organisms are sexual, hence adapt both mutation and recombination as part of their search procedures in making natural livings. But my own and other research demonstrates that recombination is essentially useless on very rugged fitness landscapes. For example, Mark Feldman and Aviv Bergman at Stanford have shown that if genes that evolve on rugged landscapes increase the frequency of recombination in model populations of organisms, they will not be selected to increase, hence establish, recombination. Yet most organisms are sexual and pay the twofold loss in fitness in requiring two parents rather than one. If so, presumably our biosphere is rife with the kinds of smooth correlated landscapes for which recombination is a good search procedure.

Then how is it that in our biosphere we should find a family of landscapes that happen to be well searched by recombination? Either such smooth landscapes are built into the physical nature of things or evolution has itself somehow brought forth the very kinds of landscapes that are well searched by mutation and recombination. Restated, assuming that mutation and recombination are, in fact, good search procedures for the kinds of fitness landscapes inhabited by we mere mortals as we were hanging around and adaptively hill climbing for the past four billion years, I ask again: Whence these wonderful fitness landscapes that are so well suited to be climbed by mutation and recombination?

Let's try another tack. Assume for the sake of discussion that I am right about my formulation of molecular autonomous agents. When life—and I argue, autonomous agents—began, their diversity was low. There are now some hundred million species, representing perhaps a thousandth of the total diversity that has wandered our globe. The rest have gone extinct. Natural games, ways of making a living, have obviously coevolved with the autonomous agents, the species, making those livings. So, as I imagined Darwin telling us in chapter 1, "The winning natural games are the games the winning species play."

Well, of course, the winning natural games are the games the winners play. But what natural games are these? The reasonable answer leaps to mind. The winning games must be those that are readily searched out by the very adaptive search procedures used by the coevolving autonomous agents themselves.

In short, a biosphere is a self-consistent coevolutionary construction of autonomous agents and ways of making a living that are themselves self-consistently well searched by the search procedures the autonomous agents are using. In colloquial terms, from our experience in economic systems, jobs come into existence with jobholders. If no one can learn or exploit a given kind of job, that sort of job will not be widely populated and will not become differentiated into a family of similar jobs.

In the biosphere, modes of making a living that are well-searched by mutation and recombination will be populated by many sibling species making their livings by playing slightly different natural games. Those natural games, therefore, proliferate. Ways of making a living that cannot be explored successfully by mutation and recombination will not afford new niches for many sibling species, so those natural games will not proliferate.

We are literally making our world together, we critters. If we couldn't make livings at it given our search procedures of mutation, recombination, and selection, we wouldn't be making our livings doing what we are doing. These comments are only the start of understanding, and I do not profess to hold much of that understanding. But I can begin to point. A biosphere is a self-consistent coevolutionary construction of autonomous agents making livings, the natural games that constitute those livings, and the search mechanisms that allow such modes of living to be persistently mastered by adaptive natural selection.

Most broadly, I believe a general biology awaits founding. And I believe that autonomous agents will prove central to that effort. The next feature of autonomous agents that I will note in closing this chapter will be central to any general biology. Precisely because an autonomous agent links exergonic and endergonic reactions in work cycles, the breakdown of high-energy sources here can be used to build up structure and organization there. Indeed, the coevolution of autonomous agents naturally leads to a linked web of exergonic and endergonic reactions within and between the autonomous agents. Breakdown of this stuff here is linked to the excess build up of that stuff there. By these linkages, sunlight spattered carelessly on this swirl of stuff ninety-three million miles from our average star cumulates into the wondrous structure of the giant redwoods, tall on the western slopes of the United States and Canada. Precisely because autonomous agents carry out work cycles, they—we—literally build a biosphere.

And the central factors underlying that buildup of organization are the same factors that apply in an economy—that merely human extension of biospheres. The central factors, in fact, center on "advantages of trade." We can see this keystone concept by supposing that you and I are the only members of a tiny economy. You begin life with an endowment of a thousand pears and a hundred apples. I begin life with an endowment of a hundred pears and a thousand apples. Suppose your happiness, or "utility," would increase if you had fewer rather than more pears and more rather than fewer apples. Alas, you have more pears than apples. I, in turn, happen to have desires such that I would be happier with rather more pears than apples. Alas, I have more apples than pears.

You and I have advantages of trade. We can both be happier if we swap some of my apples for some of your pears. It is essential to understand that, indeed, both of us can be better off by trading. Advantages of trade are the fundamental factor driving trade itself in an economy. In an actual simple economic model, advantages of trade are studied in an "Edgeworth box." Edgeworth invented a two-dimensional box representation of values, or "worths," plotted along the edge of his box (Figure 3.6). In the Edgeworth box, I am represented at the bottom-left corner, you are represented at the top-right corner. A family of equal happiness, or "isoutility" curves, show your "isohappiness" trade-offs of apples and pears at any total abundance to you of apples and pears. You are, in general, happier the more total apples and pears you have. Your happiness landscape increases from low to high like a cone-shaped mountain whose peak is located over my head. On that peak, you have all the apples and pears in the system.

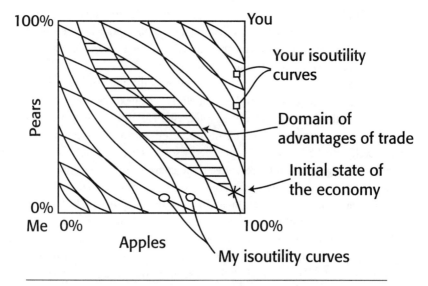

FIGURE 3.6 Edgeworth box showing advantages of trade in a pure exchange economy with apples and pears. In the initial state of the economy, I have most of the apples and you have the rest, while you have most of the pears and I have the rest. The curved lines represent my isoutility or isohappiness trade-off curves of apples for pears (lines with circles) and your isoutility curves to trade off some of your pears for apples (lines with squares).

In general, I am happier with a higher total of both apples and pears, so my isoutility curves climb upward from the corner labeled "me." Your overall happiness is greater if you have a higher total of both apples and pears, so your isoutility curves climb upward from the corner labeled, "you." The hatched region bounded by your and my isoutility curves that cross at the initial state of the economy, shows the region where we both are better off, or have advantages of trade, by exchanging some of your pears for some of my apples. For each of my isoutility curves within the hatched area, one of your isoutility curves is just tangent to my curve. The line of tangency points that traverses the hatched domain correspond to the "contract curve." If we are off the contract curve, we can both be better off by trading. On the contract curve, you cannot be better off without my being worse off. The contract curve is Pareto-efficient.

My isohappiness trade-off curves begin low at the lower-left corner and mount to a peak over your head in the upper-right corner, when I would have all the apples and pears.

The curvature of my isohappiness curves and your isohappiness curves are bent such that they are convex from my and your points of view. Therefore, if the initial economy starts with you having most of the pears and I most of the apples, as shown as a point toward the lower-right of the Edgeworth box in Figure 3.6, then that initial point of the economy lies on the intersection of a specific isohappiness curve for you and an isohappiness curve for me.

And now we can see advantages of trade. Any point that lies inside the region bounded by our two isohappiness curves is, therefore, higher on your happiness landscape and also higher on my happiness landscape. Thus, anywhere inside the region bounded by our two isohappiness curves, we are both happier. We have advantages of trade within this area bounded by our two isohappiness curves.

A few more points from Economics 100. Consider the family of your isohappiness curves and the family of my isohappiness curves. Pick one of your isohappiness curves. There will be exactly one of my isohappiness curves that just touches your isohappiness curve at a single position, thus one point of tangency. Therefore, for each of your isohappiness curves, there is a unique point of tangency with one of my isohappiness curves. Therefore we can draw a line connecting those points of tangency. In particular, we can draw a line of those tangencies across the two isohappiness curves, yours and mine, that meet at the initial apple-pear distributions to you and me at the outset, before trading, and define the region where we have advantages of trade.

The line of tangency is called the "contract curve." Along the contract curve, there is no way to exchange apples and pears that increases both our happiness. If you are happier, I am less happy. The contract curve is said to be "Pareto-efficient." There is no way to make you happier without making me less happy, and vice versa. In contrast, if we have not yet attained the contract curve, there are further advantages of trade that we can attain. The economic concept of "price" in this context is just the ratio of exchange between you and me, apples for pears. Evidently, if we attain one of the points on the contract curve, that corresponds to some exchange ratio and is the price of apples for pears.

Now, nothing in a one-shot exchange economy picks any particular point on the contract curve. We tussle along the contract curve, each trying to get all the advantages of the trade. But what if we could take our happiness, now call it utility or wealth, and reinvest it in making orchards that grew apples and pears? In an economy with reinvestment, what happens if we can take our advantages of trade and reinvest any excess so that we can create more apples and pears than we had to start with?

Then let me draw an analogy for bacterial species or other autonomous agents.

Let happiness, or the economist's utility, become "rate of reproduction," hence, fitness. Let increased happiness become "increased rate of reproduction," hence, increased fitness. Let the advantages of trade map into mutualistic interactions in which you and I, two species of autonomous agents, help one another reproduce more rapidly. Case in point: Legume root systems with microrhizzae and symbiotic fungi, in which the root and its plant capture sunlight and water and carbon dioxide and supply sugars to the fungi, while the fungi capture nitrogen from the air and fix it into amino acids and supply amino acids to the plant. Plant and fungi feed one another.

Two mutualists, A and B, can have advantages of trade. Molecules created at metabolic cost in A and secreted can help B reproduce faster. Molecules created at metabolic cost in B and secreted can help A reproduce faster. If the help is larger than the metabolic cost in both directions, both win by helping the other. Indeed, you can quickly intuit that, since both A and B will reproduce exponentially, there might be a fixed ratio of the abundance of A and abundance of B species such that each helps the other optimally. If so, then the enhancement in the growth of A and B by their mutual interaction must be the same, otherwise, either A or B would soon be exponentially more abundant than the other, and the mutual help society would fall apart.

Peter Schuster and Peter Stadler of the University of Vienna and I at the Santa Fe Institute several years ago created a simple model of two replicating RNA species, A and B, that did help one another in just these ways, and it confirmed that in the appropriate mutual help regime, the A + B mixed community outgrew A alone or B alone. Further, the growth was such that the ratio of A and B remained fixed. Therefore, the exchange of A's product molecules and B's product molecules also remained fixed at a specific point on the contract curve. That point corresponds to price. So in at least some simple models, when autonomous agents form a mutualism, A and B helping one another, they have found a means to create advantages of trade, and they can find and remain on a fixed point on the contract curve that establishes an exchange ratio—the price.

And note with the plant root and the fungi, thermodynamic work has been done by the plant to synthesize the sugars from sunlight, water, and carbon dioxide, and thermodynamic work has been done by the fungi to fix nitrogen from the air into amino acids. In a real biosphere, the linking of exergonic to endergonic reactions by which thermodynamic work is done to build up complex organization is, in fact, inextricably linked with the emergence of new advantages of trade—new, enhanced ways to make livings in new niches. In the present case, the exchange of sugar and amino acids helps both plant and fungi reproduce more rapidly.

So, as noted earlier, the fact that autonomous agents do link exergonic and endergonic reactions is central to the creation of advantages of trade and hence, new niches, new mutualistic opportunities. They lead to the vast web of an ecosystem trapping sunlight; gobbling some water, nitrogen, carbon dioxide, and a few other simple molecular species; and literally building up the vast profusion of Darwin's tangled bank. Ultimately, we should be able to build a theory that accounts for the distribution of advantages of trade, the distribution of residence times of energy stored in different forms in an ecosystem, as well as the statistical patterns of linking of exergonic and endergonic reactions in a biosphere as it builds itself and persistently explores novel ways of making a living, the novel niches that permit the success of Darwin's minor variations creating novel species for those niches.

The curious thing about evolution is that everyone thinks he understands it? Not me. Not yet. Yet I hope there may be general principles governing the self-consistent construction of any biosphere. In later chapters I will hazard a hunch or two about such general laws, but we are only at the beginning of a general biology.

Chapter 4

PROPAGATING
ORGANIZATION

T HIS BOOK, with its curious title, *Investigations*, seeks new questions about the universe. It is not always that everything is hidden and science must ferret out the mysteries by scouring for unknown facts, although often science proceeds in the manner of finding new facts. Rather it can be the case that the world is bluntly in front of us, but we lack the questions of the world that would allow us to see. There are stories, perhaps merely stories, of the response to early Spanish ships in the Caribbean by native inhabitants. The ships were not seen—there was no concept for them.

Bluntly in front of us: The closure of catalytic and work tasks in an autonomous agent by which it genuinely constructs a rough second copy from small building blocks by adroit linking of exergonic and endergonic processes. A cell, or colony of cells, is propagating this organization of process.

My aim in the current chapter is to begin to investigate what we might mean, and hence see, by propagating organization. No easy journey, this. I will begin with Maxwell's demon and why measurement of a system only pays in a nonequilibrium setting. In a nonequilibrium setting, the measurements can be stored and used to extract work from the measured system. Maxwell's demon is the clearest place in physics where matter, energy, and information come together. Yet, we will find the demon and his efforts at measurement tantalizingly incomplete: You see, only some features of a nonequilibrium system, if measured, reveal displacements from equilibrium from which work can, in principle, be extracted. Other features, even if measured, are useless for detecting such energy sources from which work can be extracted. Thus, whatever the demon's efforts, there remain the issues of just what features of a nonequilibrium system the demon must measure such that work can be extracted, how the demon knows to measure those features rather than other useless features, and how, once measured, couplings come into existence in the universe that actually extract work. Not good enough, I shall say, to assert that in principle, work can be extracted. How does work come to be extracted?

A simple example of a device that detects displacements from equilibrium and extracts work is a windmill. The vane on the windmill in effect measures the direction of the wind and pivots the windmill such that its fan blades are perpendicular to the wind. In turn, the wind does work on the blades, causing the windmill to rotate. The system as a whole measures a deviation from equilibrium (here, the direction of the wind), orients the entire system such that extraction of work by the wind is possible for the device, and it actually extracts work. The windmill turns.

The universe as a whole—from galaxies to planetary systems, and certainly our and any other biospheres—is filled with entities that measure displacements from equilibrium that are sources of energy, those entities actually do extract work. Think of the teeming busyness of a coevolving mixed microbial community of long ago, successfully linking exergonic and endergonic reactions fired by the sun and other high-energy sources. That community measured displacements from equilibrium, extracted work, and inhabited Manhattan three billion years ago, literally building high-rise microbial mat ecosystems. Its microbial descendants are constructing similar high-rise structures in the Sea of Cortez and on the Great Barrier Reef of Australia today.

Where did all this come from, this measuring of useful displacements from equilibrium from which work can be extracted, the devices coupling to such measurements, and the extraction of work used to build up new kinds of devices that measure new kinds of displacements from equilibrium to extract work in new ways? Yet a biosphere, actually constructing itself up from sunlight, water, and a small diversity of chemical compounds, does all this over evolutionary time. The biosphere does achieve persistent measuring of displacements from equilibrium from which work can be extracted and does discover "devices" to couple to those energy sources such that work can be extracted.

And since the biosphere does this, and the biosphere is part of the universe, then the universe does it. This coming into existence of self-constructing ecosystems must, somehow, be physics. Thus, it is important that we have no theories for these issues in current physics. The stark fact that a biosphere builds up this astounding complexity and diversity suggests that our current physics is missing something fundamental. A biosphere becomes complex, the universe becomes complex. I will argue that the very diversity and complexity of a biosphere begets its further diversification and complexification. I strongly suspect that the same is true of the universe as a whole. The universe's very diversity and complexity begets its further diversification and complexification.

After exploring Maxwell's demon, I will ask a physicist's question, What is work? Physicists have an answer—work is force acting through distance—given by a single number, or scalar, representing the sum of the force acting through the distance. But it will turn out that in any specific case of work, the specific process is organized in some specific way. Work is more than force acting through distance; it is, in fact, the constrained release of energy, the release of energy into a small number of degrees of freedom. It is the constraints themselves—with, as Phil Anderson points out, a kind of rigidity—that largely constitute the organization of the process. But—and here will be the hook—in many cases it takes work to construct the constraints themselves. So we will come to a terribly important circle, work is the constrained release of energy, but it often takes work to construct the constraints.

A conceptual cluster lies at the heart of the mystery. The cluster concerns the progressive emergence of organization in the evolution of the physical universe and of a biosphere. That emerging organization concerns the appearance in the evolving universe of entities measuring relevant rather than nonrelevant properties of nonequilibrium systems, by which they identify sources of energy that can perform work. Then physical entities appear that construct constraints on and couplings to the release of the identified source of energy whereby the energy is actually released and work comes to be performed. Such work often comes to be used to construct further detectors of energy sources and entities that harbor constraints on the release of energy, which when released constitutes work that constructs still further sources of energy and constraints on its release. It should be clear that we have at present no theories about these matters, nor even a clear concept of the subject matter of such theories.

The heart of the mystery concerns a proper understanding of "organization" and "propagating, diversifying organization." Most profoundly, the mystery concerns the historical appearance since the big bang of connected structures of matter, energy, and processes by which an increasing diversity of kinds of matter, sources of energy, and types of processes come into existence in a biosphere, or in the universe itself. This is what lies directly before us but which we have not been able to see. A biosphere does all the above. Ours has for four billion years of awesome, ill-understood creativity. Doubt it? Open your eyes and look around you.

The universe, since the big bang, was and remains out of equilibrium, or vastly nonequilibrium. It was a profound insight in the development of equilibrium thermodynamics to recognize that the energy present in the thermal motions of an equilibrium gas system could not be extracted to do work. But we might ask a similar question of the nearly featureless, profound *nonequilibrium* of the early universe. How, in the absence of specific structures and processes, could the nonequilibrium universe couple that enormous energy to the specific generation of anything at all? Part of the answer lies in the concept of broken symmetries. Consider a pole standing vertically on a horizontal plane. In due course, it will fall over under the influence of gravity. Prior to falling, its range of possible directions to fall is the full circle. After it falls, it points in some specific direction. By falling, the pole has broken the circular symmetry of the system and come to a specific orientation. Thus part of the answer to the emergence of specific structures lies in the expansion and cooling of the universe, with the associated sequences of symmetry breakings that split the four fundamental forces, yielded a quark-gluon soup that cooled into other elementary particles, then atoms, simple molecules, self-gravitating masses, galaxies, giant molecular clouds, and second-generation stars.

As symmetries broke, the variety of matter and process increased. As the variety increased, the pairwise diversity of matter and processes increased roughly as the square of the diversity. Hence, it became more probable that specific pairs of spontaneous and nonspontaneous processes might become linked in a variety of ways, capturing the energy resources of the spontaneous processes that could then flow in constrained ways into the nonspontaneous processes to yield novel consequences. Among those consequences are the construction of new structures able to measure sources of energy. Among the other consequences are the generation of novel and specific nonequilibrium energy sources and of structures and constraints that might couple to those novel specific sources of energy. The couplings and constraints, in turn, channel the release of energy in specific ways that constitutes the work that is done to construct still further novel energy sources, measuring structures, couplings, and constraints. This, in a nutshell, is the universe diversifying, constructing structures and processes, propagating and elaborating wondrous organization.

In chapter 2 I introduced the chemical adjacent possible and will return to it in later chapters. In terms of molecular diversity and other types of diversity, the universe and the biosphere keep advancing into a persistent adjacent possible. New kinds of molecules with new properties themselves and in couplings with other kinds of molecules persistently arise on planet Earth, and presumably in the giant cold molecular clouds that are the birthplaces of stars in most spiral galaxies. The new species of molecules afford the novel exergonic and endergonic reactions, novel constraints, and novel sources of energy that are part of the creativity outside our collective window.

Yet we hardly know how to say what this propagation and elaboration of organization and process is, nor have we a clue about whether there may be general laws that govern such self-constructing nonequilibrium processes. Such a law could be my hoped-for fourth law of thermodynamics for open self-constructing systems.

We have begun with autonomous agents. But we are here driven beyond biospheres. What are the general conditions that allow such self-constructing nonequilibrium processes to flourish? Are biospheres the only examples? What of the evolution of the geology of a planet, a solar system, a galaxy, the universe as a whole? Are there ways of thinking about the emergence of structures that measure and discover sources of energy in nonequilibrium systems, together with the emergence of structures and processes that couple to sources of energy, do work to construct constraints, and propagate the constrained release of the discovered energy such that more diverse structures, constraints, and processes can arise, de novo, in the adjacent possible of the evolving universe?

Is the universe highly diverse, and is our biosphere diverse, because there is some general law or tendency for such nonequilibrium self-constructing systems to diversify? I confess I suspect so. In an intuitive nutshell, in a nonequilibrium setting, the greater the diversity of structures, potential reactions, or other transformations among structures, measurement processes and devices, coupling devices, and constraints that already exist in a ramified web of propagating structures, reactions, work, measurement, constraint and coupling constructions, the easier it is for the total system to generate new kinds of molecules or other structures, processes, measurement devices, couplings, and constraints such that a biosphere or the universe can expand into the newness of its adjacent possible. But those new structures, processes, measuring devices, couplings, and constraints in turn increase the total diversity, hence, enable yet further expansion into the adjacent possible, creating perpetual autocatalytic novelty on timescales that must be vastly longer than the current age of the universe.

The universe, in short, is breaking symmetries all the time by generating such novelties, creating distinctive molecules or other forms which had never existed before. Indeed, there may be a general law for biospheres and perhaps even the universe as a whole along the following lines. A candidate fourth law: As an average trend, biospheres and the universe create novelty and diversity as fast as they can manage to do so without destroying the accumulated propagating organization that is the basis and nexus from which further novelty is discovered and incorporated into the propagating organization.

Autonomous agents themselves, self-reproducing systems carrying out one or more work cycles linking exergonic and endergonic processes in a cyclic fashion that propagate the union of catalysis, constraint construction, and process organization that constitute that autonomous agents are but the most miraculously diversifying examples of this universal process in our unfolding, ever-changing universe.

Maxwell's Demon

Arguably James Clerk Maxwell was the greatest scientist of the nineteenth century, notwithstanding giants such as Carnot, Boltzmann, and Darwin. While his most radical work is captured in the Maxwell equations for electromagnetic fields, which introduced the fundamental concept of fields into physics, Maxwell concerned himself deeply with the puzzle Carnot had raised in what is now called the second law of thermodynamics.

Consider again a thermodynamically isolated system. That is, consider some box containing a gas, isolated from any change in its energy or mass arriving from the outside. There are N gas particles in the box, and as noted earlier, we can consider the positions and momenta of all N particles. Each position and each momentum can be decomposed into three numbers defining position and motion in the three spatial directions. Hence, the entire state of the N particles of gas can be defined by $6N$ numbers, plus a specification of the interior boundaries of the box.

As described above, all the possible states of this $6N$ system of particles can be divided into very small volumes of states, which we will call microstates. Again, as noted in chapter 3, a macrostate is a collection of microstates. In particular, the equilibrium macrostate is a collection of microstates having the property that the gas particles are nearly uniformly distributed in the box, with a characteristic equilibrium distribution of velocities that Maxwell himself worked out. This equilibrium macrostate has the further important properties that (1) vastly many microstates are in the equilibrium macrostate; (2) a few macroscopic features—temperature, pressure, and volume—suffice to specify the equilibrium macrostate.

In terms of microstates and macrostates, as we saw, the second law can be reformulated in its famous statistical mechanics incarnation. The second law becomes the statement that, at equilibrium, the system will flow from any initial macrostate such that it spends most of its time in the equilibrium macrostate. This statement of the second law does not preclude the extremely improbable case in which the N particles just happen to flow to one corner of the box. Thus, the second law is a statistical law in statistical mechanics.

But now Maxwell enters and invents a "wee creature," later dubbed Maxwell's demon. (I confess that I find the use of the term "demon" here more than slightly interesting. Maxwell's demon is almost an autonomous agent. While the demon is not defined as I have done, you will soon see that he seems to be able to make decisions and to act on the physical world. I suspect it is more than a mere coincidence that Maxwell and we seem forced to use this kind of intentional language. In fact, an odd feature of physics is that experimenters, who are outside the "system," are always busy intentionally setting up experiments and preparing quantum systems in desired states. Surely, in a full theory the experimenters themselves, each an autonomous agent, would be part of the theory? And if not, why not? In chapter 6 I return to this theme, for it relates to our incapacity to finitely prestate the configuration space of a biosphere.)

Maxwell asks us to consider the very same box with N particles in it. But he imagines the box to be divided into two chambers by a wall with a window in it. In the window is a flap valve. When the flap valve is open, gas particles can pass from the left to the right box via the window, or from the right to the left box via the window.

Now, smiles Maxwell, suppose the initial state of the gas in the box is in the equilibrium macrostate. No macroscopic work can be done by the equilibrium system. That was Carnot's central point. There is plenty of energy in the random motions of the gas particles, but there is no means to extract mechanical work from it, say, to drive a piston. Next, says Maxwell, warming to his point, "Imagine that my wee friend operates the flap valve such that, whenever a fast gas particle approaches the window from inside the left box toward the right box, he opens the flap and lets the faster than average, hence hotter, gas particle through. And suppose my demon also operates the flap value to let the slower than average, hence cooler, gas particles pass from the right to the left box. Well, soon the left box will be cool and the right box will be hot. And now," concludes Maxwell with a broad smile, "we can use the macroscopic temperature difference between the left and right boxes to extract mechanical work, say, by driving a piston."

There you have it. Maxwell posed a severe question for statistical mechanics and the second law. It appeared that the actions of the demon might circumvent the second law.

Maxwell's demon has set a puzzle that is still not fully resolved. An important step in "saving" the second law was taken by Leo Szilard, who also conceived of the nuclear chain reaction one day in London and helped set in motion the development of the atomic bomb and atomic energy. Szilard carried out a calculation linking, for the first time, the concept of entropy and a new concept of information. The "entropy" of a system is a measure of its disorder. Recall that we can define the volumes of different macrostates by the numbers of microstates each macrostate contains. For convenience, take the logarithm of the number of microstates in each macrostate. In addition, each macrostate also has a probability of being "occupied" by the system. Multiply the logarithm of the number of microstates per macrostate by the probability that the system is in that macrostate. Now add up all these quantities for all the macrostates. The total is the entropy of the system.

Statistically, of course, the entropy of a system either increases over time or is constant. At equilibrium it is constant. If the system is released from an initially improbable macrostate, its initial entropy for the first period of time is low since most macrostates are not occupied. However, over time it will tend to spread out over all possibilities, and the sum of the probabilities of occupancy times macrostate volumes will increase to the equilibrium value.

Szilard took a first step in thinking about what Shannon later called information. Roughly Szilard realized that when the demon lets a faster or a slower gas particle pass specifically into the left or right box, respectively, then the total entropy of the system is decreasing a little bit. But in turn Szilard estimated the amount of work that must be done by the demon to discriminate that the gas particle is faster or slower than average. It turns out that the work that must be done, hence the energy utilized, equals the work that can later be extracted from the system after the fast and slow particles are separated into the two boxes. Since the work done by the demon equals the work that later can be extracted from the system, no net work can be extracted from the equilibrium system, and the second law is saved.

The link to information due to Shannon comes next. Shannon was concerned with transmitting signals down wires. He brilliantly thought of the minimal signal as a yes or no answer, hence representable as the binary 1 or 0, now called a "bit." Shannon considered the entropy of a source sending a prospective signal as the set of possible messages that might be sent, where each message was to be weighted by the probability of actually being sent. He thought of receiving a message as reducing the entropy, or uncertainty, about which message was actually sent, given the initial set of possible messages. Thus, Shannon wound up reinventing the same mathematics that covers entropy. Here there is an ensemble of messages, and each can be thought of as occupying a volume in a space of possible messages. Each message is sent with some probability from the source. So Shannon took the logarithm of the volume in message space occupied by a message and multiplied it by the probability that that message was sent from the source. If the fraction of the total volume of message space occupied by a given message is "p," then the logarithm of this volume is "$\log p$" and the probability of that volume is "p." Thus, the logarithm of a probability of a message multiplied by that probability itself is "$p \log p$." The sum of these "$p \log p$" terms for the total set of messages at the source is the entropy of the source. Reception of a signal reduces the receiver's uncertainty about what is being sent from the source, hence is a negative entropy. Shannon's information measure is, thus, just the negative of the normal entropy measure.

The link established by Szilard between information and Maxwell's demon is, roughly, that the discrimination by the demon that a given gas molecule is faster or slower than average and whether it is coming from the left or right box (hence, whether he should open or close the flap valve) constitutes a measurement that extracts information about the gas system, hence, lowers the uncertainty about the gas system, hence, lowers the entropy of the gas system.

Importantly, there is an implied observer in discussions about entropy. Thus, a physicist might typically say that the entropy of a system is due to "our coarse graining" of the system into (arbitrarily) chosen macrostates. If "we" had more information about the microscopic states of the system, our more refined coarse graining would reduce the entropy of the system from our point of view. Indeed, there has been some genuine confusion about the role of the observer and his more or less arbitrary choice of coarse graining in the concept of entropy.

One resolution to this confusion has been suggested by Rolf Sinclair and Wojciech Zurek, who have returned to the demon problem with a wonderful set of concepts. When the demon has at it with the flap valve, he is, in fact, performing measurements on the gas system. As he performs the measurements, he "knows" more about the detailed state of the system. Now just what might it mean to know about the gas system? One useful sense of "know" is that the demon has some compact description of the state of the gas system. Indeed, the compact description of the equilibrium state is about as compact as you can get: A few macroscopic variables—temperature, pressure, volume—suffice.

One modern sense of a compact description of something is a computer program. We are to think of the computer program as a calculating engine. We give it initial input data. It has some program, typically written as a sequence of binary numbers, 0 and 1, and the program operates on the input data, also a string of binary symbols, and churns out an answer. Then the concept of a compact description becomes the concept of the shortness of the symbol string giving the input data and the shortness of the program. In order to maximize compression, we must get all redundancy out of both the input symbol string and the symbol string representing the program.

Sinclair and Zurek have independently carried out work that shows the following: Initially, as the demon operates, his knowledge about the system increases, hence, the entropy of the gas system decreases. But at the same time, as the demon's information about the system increases, the length of the most compact description of the system increases as well. In fact, the length of the most compact description increases, on average, exactly as fast as does the decrease in the entropy of the gas system.

But as the length of the most compact description increases, bit by actual bit, its information content increases, bit by bit. Thus, for each bit in reduction of the entropy of the gas system achieved by our measurements, the information content of the most compact description increases, on average, exactly as rapidly. Or, as Zurek says, in the modern interpretation, the sum of the entropy of the gas system plus the observer's knowledge about that system is a constant for an equilibrium gas system.

Well, we could still cheat and extract work from our measured gas system using the information about its microstate achieved by all the measurements. But Sinclair notes that, in the long run, the cheat will not work. We have had to record the information about the gas system somewhere, say, in the registers on a silicon chip. At some point in a closed system, the chip will be filled up with bits in registers. To keep measuring the equilibrium system, we will have to erase the chip. And Sinclair did the calculation that mirrors Szilard's. To erase a memory-stored bit has a minimal energy cost that exactly balances the work we could get from the gas system by using the stored information about the system. The second law, again in the statistical sense, holds. No macroscopic work can be done by an equilibrium system. Measurement does not pay in an equilibrium setting.

Why this long discourse? Because it does pay to measure the gas system if the gas system is not at equilibrium. Think of a simple example: The gas particles in the left box are actually hotter than the gas particles in the right box. Thus, pressure in the left box is higher than in the right box. If the flap valve is opened, gas will tend to flow from the left to the right box until equilibrium is established. Note that a very simple, compact description has captured these features of the nonequilibrium system, and work can be extracted as the gas system flows to equilibrium.

More generally, Zurek's point is that as measurements are performed on a nonequilibrium gas system, the length of the most compact description increases more slowly than the knowledge thus gained reduces the entropy of the system. It pays to measure the nonequilibrium system in the sense that those measurements specify the displacements from equilibrium that constitute energy sources that can be utilized to extract work.

So the demon is indeed a place in physics where matter, energy, information, and indeed, work, come together.

Let's consider just how work might actually be extracted in the classical Maxwell demon situation with an ideal gas in two boxes separated by a partition with the window and flap valve. As a simple example, consider again the tiny windmill mentioned above, consisting of a fan and a vane perpendicular to the fan. Let the windmill be located very near the window with the flap valve inside the total gas system. If the flap valve is opened, a wind will pass transiently from the left to the right box. The windmill's vane will measure the direction of the wind and actually orient the windmill fan blades perpendicular to the wind. The wind will cause the fan to turn, thus the turning fan extracts mechanical work from the system until equilibrium is reached.

But now we need to pause and reflect, for the story of the demon is both tantalizing and incomplete. Consider again our tiny windmill. What feature of the total gas system was measured and detected such that work could be extracted? Roughly, the wind from the left to the right box.

But not all measurements of the two-box system would have resulted in information that was useful, in the sense that work could have been extracted by the actual box in its actual configuration. For example, the box with the flap valve separates the left and right boxes; suppose that there is an identical number of gas molecules in the two equal-sized boxes and that the gas in the left box is hotter than the gas in the right box. Further, suppose the demon measures the number and instantaneous locations of all the gas particles in the left and right boxes. The fact that the particles in the left box are hotter than those in the right box, hence are moving faster than those in the right box, would not be revealed by a measurement of the instantaneous numbers and locations of all the gas particles in the left and right boxes. To measure faster motion, the demon must measure positions at two time moments or some other feature, such as the recoil of the box's walls from the momentum transferred by the hotter versus cooler gas particles in the left and right boxes as they bounce off the wall. So, just how does the demon decide

(Figure 4.1) or come to measure the relevant properties such that an energy source is successfully identified such that work can be extracted?

We have, in fact, no answer as yet.

But this is an essential issue. Only certain features of a nonequilibrium system will, upon measurement, reveal a displacement from equilibrium that can actually be used to extract work. Other features, if measured, are useless with respect to revealing a displacement from equilibrium that can be used to extract work by any given specific system.

It is important to stress that we have here a sense of "useful" outside the context of autonomous agents. Useful measurements detect features of displacements from equilibrium that reveal energy sources from which work can be extracted. Only some measurements are actually useful in this sense in a biosphere, a geosphere, or a galaxy. These useful measurements participate together with the coming into existence of devices that extract work used to build further measurement and work

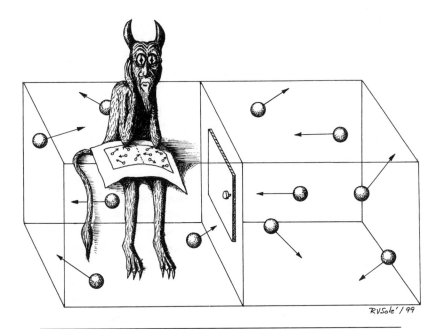

FIGURE 4.1 Maxwell's demon flummoxed. The demon is considering the gas system in which he finds himself and is trying to decide what feature of the system to measure in order to detect if the system is displaced from equilibrium in some way such that work can be extracted. Unfortunately, he decides to measure the positions of all gas particles at a single instant, a strenuous measurement from which he cannot detect a displacement from equilibrium from which work can, in principle, be extracted. How does the demon know what to measure? (Figure drawn by R. Solé)

extraction structures, in the gradual buildup of the diversity of structures and processes of a biosphere, a geosphere, a galaxy, or a universe. This buildup is part of why the universe is complex.

I believe that we can ultimately create a statistical theory of the probability of the generation of specific novel processes, structures, and energy sources; propagation of measurements; detection of useful sources of energy; and couplings of structures and processes to the energy sources to extract work and progressively build up still further new structures, energy sources, and processes—all as a function of the current diversity of structures, transformation processes, and measuring and coupling entities. Such statistical theories should be constructable, for example, for a giant cold molecular galactic cloud or early prebiotic planet or, most fundamentally, the expanding universe as a whole. We need a theory in which symmetry breaking begets further symmetry breaking in a progressive construction of diversifying structures and processes. Chapter 2, with its discussion of the origin of self-reproducing molecular systems as a phase transition to supracritical behavior in catalyzed chemical reaction graphs as a function of molecular diversity and the ratio of reactions to molecular species, is a partial prototype for such a statistical theory. A further partial prototype is present in chapter 3, with its discussion of autonomous agents as self-reproducing physical systems that do successfully measure displacements from equilibrium and do successfully evolve to couple exergonic and endergonic reactions to achieve completed work cycles. The vast and richly coupled network of coupled exergonic and endergonic reactions in the global ecosystem is proof positive of such propagating construction in the physical universe. In chapter 10 I will discuss a quantum analogue to such a theory, in which complex quantum systems that couple tend to "decohere" irreversibly to classical behavior and thereby progressively build up complex classical structures.

It is also important to unpack the sense, three paragraphs above, of "actually" and "any specific system." Consider a single gas particle in a box. Measure its location, left or right of any arbitrary surface transecting the box. Here "arbitrary" means that we can choose to perform any such measurement we wish by placing the partition arbitrarily in the box. If we know the particle is to the left of a given arbitrary partition, we can in principle extract work by allowing the particle to pass through a window in the partition and do work on a fan as it passes to the right box. Hence, it seems that in principle any such arbitrary measurement can detect a source of energy that can be used to extract work.

But the conclusion is false that any arbitrary measurement of our single-gas-molecule system can detect a displacement from equilibrium from which work can be extracted. The "in principle" just above includes the idea that, having made an arbitrary choice of placement of the partition and a measurement of which side of the partition the particle is in and, hence, having detected by that arbitrary measurement the displacement from equilibrium that is a source of energy, we can *af-*

terward decide on a construction procedure that will utilize the information about the displacement from equilibrium to extract work from the measured, nonequilibrium system. In short, we can place the windmill in the system after we have measured the location of the gas particle. We measure first, then place the windmill in the compartment that does not have the particle of gas, such that that particle, upon passing through the flap valve, will cause the windmill to turn slightly.

But what if we already have constructed the system that is to extract the work, as in the tiny windmill case, and already mounted the windmill at a specific location inside the box? Thereafter we perform an arbitrary measurement by placing the partition in the box and then locate the gas particle. We may have placed the partition in the box such that the windmill is on the same side of the partition that has the gas molecule, rather than placing the partition such that the prepositioned windmill is in the empty side. No net work can be extracted. The gas molecule will repeatedly bounce off the windmill fan from all angles. No net rotation of the fan can occur.

Thus, in a concrete context, when we can no longer alter the work-extracting structure, such as the location of the windmill, but perform the measurement after the work-extracting system is in place, then only certain measurements of the nonequilibrium system will detect sources of energy that can couple to the work-extracting structure such that work is extracted. Other measurements of the extant nonequilibrium system may be utterly useless in the sense that no sources of energy that can couple to the work-extracting system are detected.

We see the hints here of something new. Only certain features of a given nonequilibrium system, if measured, will result in detection of sources of energy that might become coupled to specific other processes that, by doing work, propagate macroscopic changes in the universe. Moreover, the tiny windmill is an example of a device that not only detects the wind from the left to the right box, but also orients the fan perpendicular to that wind and has couplings and constraints embodied in its structure such that mechanical work is actually extracted.

Fine, but we built the tiny windmill. How do such coupling structures that link identified sources of energy to the carrying out of work come to exist in the universe on their own? There is not the slightest doubt, for example, that such entities have come into existence in our biosphere as autonomous agents have coevolved over the eons. Thus, a host of new questions are raised. In the beginning, presumably, the universe was simple, homogeneous, featureless, almost isotropic. Now it is vastly complex. In the beginning, the early Earth had a paucity of complex molecules, chemical reactions, linked structures and processes. Now it is vastly complex.

The universe as a whole has witnessed the coming into existence of novel structures and processes; so too has the biosphere. Where no difference existed, differences have come into existence. In a general sense, the persistent emergence of different structures and processes is the persistent breaking of the symmetry of the universe. What feeds this apparent propagating diversity? One aspect may be the following. Consider again the case of the box with the flap valve and something simpler than a fan, say a small mica flake suspended in the cooler of the left and right boxes. If the flap valve "be opened," a wind from the hotter to the cooler box is transiently present. This is a simple displacement from equilibrium, and a simple device, the mica flake, will be made to quake, hence, extract mechanical work.

Now consider an antiferromagnetic material. Such material has magnetic dipoles that, when adjacent, prefer to point in opposite directions. The north pole of one prefers to be adjacent to the south poles of its neighbors. If arranged along a straight line, an antiferromagnetic material has two equivalent lowest-energy "ground" states, NSNSNSNSN versus SNSNSNSNS. Now consider a subtle displacement from one of these lowest-energy states, say NNNSNSSSN. Here, rather than alternating N and S poles being next to one another, runs of NNN and SSS occur. The energy of the total system would be lowered if the dipoles flipped orientation to come closer to one or the other of the ground energy states. Therefore, at a sufficiently low temperature such that the system can flow to and remain at a ground state, the NNNSNSSSN antiferromagnet is displaced from its lowest-energy equilibrium state, and in principle, work could be extracted from this system as it relaxes to one of the two lowest-energy states. But notice now that, compared to detecting the direction of the wind by the mica flake, a rather complex and subtle measurement must be made by any measuring device that is to detect the subtle displacement from equilibrium and that any device that is to use that displacement to extract work must be correspondingly subtle. Roughly speaking, a measuring device must be of similar complexity to the antiferromagnet. Indeed, a second antiferromagnet could serve as a measuring device if it were near its own ground state and brought into proximity to the first antiferromagnet. The runs of SSS and NNN in the first antiferromagnet, brought close to a second one with ground state runs of SNS and NSN could cause the first antiferromagnet to flip closer to its ground states. Hence, the measuring-detecting-extracting device must be more structurally and functionally complex than a mere mica flake considered as a thin planar crystal.

The linked exergonic and endergonic organic chemistry reactions present in the molecular autonomous agents that we call cells exemplify just this structural and functional subtlety. The electric charge distribution on two complex organic molecules brought into proximity, coupled with the modes of translational, vibrational, and rotational motions, constitute the subtle means to measure displacements from equilibrium, couple to those displacements, and achieve linked catalyzed exergonic and endergonic reactions. As the molecular diversity of the biosphere increases, more such molecular species displaced from equilibrium come into existence, more such molecular species able to detect such displacements from equilibrium come into existence, more such coupled catalyzed exergonic and endergonic reactions come into existence.

In general, it would begin to appear that as a higher diversity of entities come into existence—entities that are then necessarily more complex—their modes of being in nonequilibrium conditions increase in diversity and subtlety. In turn, the very existence of sets of these increasingly diverse and complex entities gives them an increased number of ways, and so an increased probability, to couple with one another such that one may measure a displacement from equilibrium of the other; hence, these entities happen upon a source of energy that can be and is extracted to do work. In turn, that work may drive nonspontaneous processes to create still more complex molecular species or other entities in the adjacent possible.

In short, there appears to be some positive relationship between the diversity and complexity of structures or processes and the diversity and complexity of the features of a nonequilibrium system, which can be detected and measured by the detecting structure to identify a source of energy, then couple to the source of energy and actually extract work. If there is a relation such that diverse and complex features of nonequilibrium systems useful as sources of energy can best be detected by equally diverse and complex structures, then there appears to be some generalized "autocatalytic" set of processes in the universe since the big bang, and in a biosphere, by which nonequilibrium systems of increasing complexity and diversity arise, provide sources of energy of increasing subtlety and complexity, and in turn are detected and extracted by the increasingly complex structures that arise.

Of course, to hint the above is to hint an initial answer. At least in our biosphere, the cumulative coevolution of autonomous agents has, in the past four billion years, achieved precisely such a diversification. Cells and organisms have achieved astonishingly ramified and subtle detectors that measure sources of energy, plus coupling devices, that extract work and use it to build rough copies of themselves. Thus, metabolism in cells is a coupled web of chemical reactions among simple, complex, and very complex organic molecules, ranging from carbon dioxide to proteins comprised of thousands of amino acids. The catalytic sites of enzymes possess high stereospecificity—that is, shape specificity—for the transition state of the substrate(s) of the reaction. Such reactions may release energy or may couple the release of energy to the endergonic synthesis of other molecular species. Cells are replete with equally stunning receptor complexes decorating their surfaces. Binding a ligand to a receptor may trigger a complicated sequence of reactions leading to the synthesis of hundreds of different molecular species. But the high specificity of molecular interactions in a cell are precise examples of the coming into existence of richly nuanced, structurally and procedurally complex molecular processes that measure and detect sources of energy, and couple those sources to the carrying out of further chemical, electrical, or mechanical work.

A coevolving biosphere achieves exactly the emergence of such self-constructing diversifying organization. Whether galaxies, planetary, stellar, or other systems do as well is an open question. Again, one senses the possibility of a statistical theory of the propagation and self-elaboration of such linked structure transformational systems.

Work

Let's turn to the concept of "work."

I have detailed evidence that work is a puzzling concept. I am deeply proud that Phil Anderson, one of the world's best physicists, is a close friend. One day over an Indian dinner in Santa Fe, thinking of the issues above and of more to come, I said, "Phil, the concept of work is rather puzzling." Phil cracked off a bit of chapati, scooped some chutney onto it, paused, and said, "Yes."

Thank God. I'm not a physicist, so I was glad to get through that hurdle.

I shall proceed in steps. First, let's just consider the physicist's definition of work as the integral of force acting through distance. The physicist has in mind something like Newton's laws, where $F = MA$. And we understand distance, plain old nonrelativistic distance. So the work done is given by just adding up little increments of the force acting on a mass and accelerating it through a distance.

But already there is a bit of a puzzle. In any specific case of work done, some direction of application of force is specified in three-dimensional space, some actual direction of motion of the mass is specified in three-dimensional space, and some actual coupling mechanism is in place such that the force does act on the mass and get it to accelerate in that direction. How does the "specification" of a direction come to be? How does the organization of the specific case of work come about?

Now in normal physics, say, college-level physics, all these specifications occur at the beginning of the problem, in the statement of the initial and boundary conditions. The billiard balls are in such and such positions on the billiard table, the cue is moved with such and such velocity and strikes a given ball in such and such a position with such and such velocity. Now, given Newton's calculus, let us compute the forward trajectory of the balls on the table. So the puzzle of where the initial and boundary conditions come from, and the specific coupling of cue to ball, are "hidden" in the initial and boundary conditions of "the problem" and in how Newton taught us to calculate. In short, the problem of the organization of the process in any specific case of work is hidden from view in the initial and boundary conditions of the usual statement of the physical problem. In effect, this choice is the choice of the "relevant" degrees of freedom, which is equivalent to the choice of the boundary conditions versus the dynamical variables of the system.

But an evolving biosphere is all about the coming into existence in the universe of the complex, diversifying ever-changing initial and boundary conditions that constitute coevolving autonomous agents, with their changing organization of capacities to measure and detect energy sources, and couple those detected energy sources to systems that sometimes extract work. We will return in a subsequent chapter to ask if it makes sense to try to finitely prestate the initial and boundary conditions of a biosphere. I will claim that it does not. I will claim that we cannot finitely prespecify the configuration space of a biosphere, hence, we cannot finitely prespecify the initial and boundary conditions of a biosphere.

If so, then we cannot hide the issue of the organization of work processes in a statement of the initial and boundary conditions of the biosphere. We must grapple with the emergence and propagation of organization itself on its own terms. If so, perhaps there is something amiss with the way Newton taught us to do science in his spectacular career.

Let's take a second look at work. Consider an isolated thermodynamic system. At equilibrium the system can do no work. But let the system be partitioned into two or more domains, say, by a membrane. Ah, then one part of the system can do work on the other part. For example, if the average pressure in one part is higher than in another part, the first part can bulge the membrane into the second part.

Where did the membrane come from? How does the system come to be partitioned? Is this just another initial or boundary condition hiding from view the question: Where did this organization of stuff and process come from? Meanwhile, note that the concept of work appears to require that the universe be partitioned. Regions of the universe must be distinguished (by what or whom?) such that work manages to happen.

Now I come to a definition I like, due to Atkins in his book on the second law. Atkins defines work as "the constrained release of energy." Work is, says Atkins, a "thing."

Think about the cylinder and piston in the idealized Carnot cycle, with the hot, compressed working gas in the chamber. What are the constraints? The cylinder and the piston, the position of the piston in the cylinder, the grease between the piston and cylinder are constraints. These roughly suffice, together with the hot gas compressed in the cylinder head, for work to happen as the hot gas expands and pushes on the piston.

Where did these constraints come from? In actual fact, in the current case some human, or some machine made by a human, did work to construct the cylinder, the piston, assemble the piston into the cylinder with working gas and grease in place. Then more work was done to compress and heat the gas by pushing on the piston from the outside.

So we appear to come to an interesting circle. It sometimes takes work to construct constraints, and it takes constraints to get work.

Does it always take work to construct constraints? No, as we will soon see. Does it often take work to construct constraints? Yes. In those cases, the work done to construct constraints is, in fact, another coupling of spontaneous and nonspontaneous processes. But this is just what we are suggesting must occur in autonomous agents. In the universe as a whole, exploding from the big bang into this vast diversity, are many of the constraints on the release of energy that have formed due to a linking of spontaneous and nonspontaneous processes? Yes. What might this be about? I'll say it again. The universe is full of sources of energy. Nonequilibrium processes and structures of increasing diversity and complexity arise that constitute sources of energy and that measure, detect, and capture those sources of energy, build new structures that constitute constraints on the release of energy, and hence drive nonspontaneous processes to create more such diversifying and novel processes, structures, and energy sources.

I find it delightful that we hardly have the concepts to state these issues; surely we have as yet no coherent theory for this burgeoning of process and structure. Whatever it is, a biosphere does it. It was quite barren in Nebraska, wherever Nebraska was, four billion years ago. Not now.

Propagating Work

By way of whimsy, consider Figures 4.2a and 4.2b. Figure 4.2a exhibits a cannon, clearly marked "cannon," firing a cannonball, clearly marked "cannonball," that hits the ground some distance away, creating a hole, clearly marked "hole." In addition to creating the hole, the cannonball, now embedded in the bottom of the hole, has created hot dirt, marked "hot dirt."

Nonpropagating Work

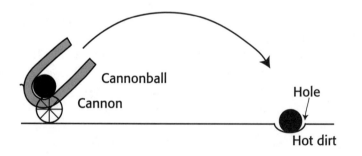

FIGURE 4.2a Nonpropagating work. A cannon fires a cannonball that hits the ground some distance away creating a hole and hot dirt.

In Figure 4.2b I exhibit a device—a Rube Goldberg device, in fact—of which I am extremely proud. The same cannon as in 4.2a now fires the same cannonball, which, however, hits a paddle on a sturdy paddle wheel I constructed. Once struck by the cannonball, the paddle wheel is set to spinning. Prior to my firing the cannonball, I contrived to tie one end of a red rope around the axle of the paddle wheel and a modest size bucket to the other end of the rope. Thereafter, I dropped

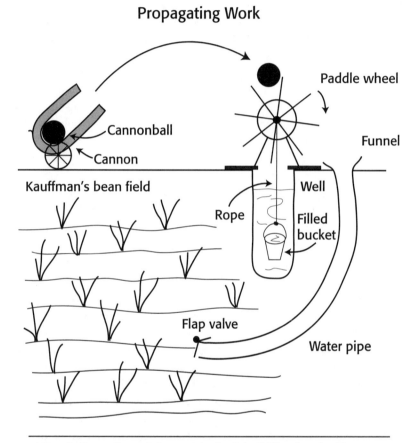

Propagating Work

FIGURE 4.2b Propagating work via my outstanding Rube Goldberg invention. The same, or identical cannon as in the previous figure, fires the same, or identical cannonball. But now, the cannonball hits the sturdy paddles of the sturdy paddle wheel I built and placed beside the well I dug. I have attached a red rope to the axle of the paddle wheel and tied the other end to a bucket, which is down in the well and filled with water. The spinning paddle wheel winds the rope up on the axle and raises the bucket, which tips over the axle, pouring water into the funnel leading to the water pipe. The flowing water opens the flap valve at the end of the pipe, thereby watering my bean field. I feel my invention is a substantial addition to agricultural technology.

the bucket down the well. The water-filled bucket has now rested, silent and wait-ing, until the cannonball strikes the paddle wheel, whereupon the wheel spins, the red rope winds up, pulling the water-filled bucket up the well, up against the axle, which tilts the bucket over—you will have to imagine this part—and pours the water into a long funnel that slopes down from the wellhead toward my bean field. When the water from the bucket arrives at the bottom of the water pipe, it pushes against a flap valve, thereby opening the valve and watering my bean field. You can see why I might be proud of my machine.

What is the difference between 4.2a and 4.2b? The point of the cannon and can-nonball in the two figures is to emphasize that there is the same total input of en-ergy into the two cases. The explosion of gunpowder is evidently the same, as is the flight of the cannonball. Obviously, in Figure 4.2a, most of the energy carried by the cannonball is dissipated as heat, random molecular motions induced in the particles of dirt. Indeed, I might have sent the cannonball bouncing along a large steel plate rather than hitting mere dirt. In the case of the plate, no hole would have formed, and hot steel would have been the consequence.

In Figure 4.2b, my Rube Goldberg device achieves a rudimentary—or sophisti-cated, depending upon pride of inventorship—propagation of macroscopic conse-quences in the universe. Note the linking of spontaneous and nonspontaneous processes—the arc of the cannonball imparting energy that winds the wheel and lifts the water-filled bucket. Note also the constraints everywhere present that coor-dinate the flow of energy into the specific, if slightly comical, unfolding of events.

In fact, my fine Rube Goldberg device does not quite demonstrate all I might wish it to show, for it does not demonstrate the use of the release of energy to actu-ally construct constraints. However, an ingenious modification of my device, of which I am also deeply proud, demonstrates constraint construction. Let us mod-ify the device such that the cannonball, after hitting the paddle wheel and setting it spinning, is deflected downward onto the ground and digs a long shallow groove in the dirt, with high sides due to the displaced dirt. Let this groove lead to the bean field and guide the water spilled from the bucket such that it flows to water the bean field. The digging of the groove in the dirt by the cannonball constitutes the construction of constraints on the release of energy, for the water flowing down the gravitational potential to the bean field is just such a constrained release of energy.

My Rube Goldberg device propagates work; it succeeds in creating a sequence of coordinated macroscopic changes in the physical universe. I do not know a for-mal definition of "propagating work," so, in the absence of anything better, I will point at what I mean by Figure 4.2b.

We have some clues in place now. Work is the constrained release of energy. Often constraints themselves are the consequence of work. I have tentatively de-fined an autonomous agent as a self-reproducing system that carries out at least one work cycle. In turn, this led us to note that an autonomous agent is necessarily a nonequilibrium device, therefore, that it stores energy. To think about work cycles, we have been driven to ask about Maxwell's demon, measurement, when and why measurement pays, thence to what features of a nonequilibrium system are measured such that they constitute a source of energy, thence to how couplings arise that capture the energy source, thence to work and constraints, and now to propagating work due to the occurrence of linked sets of constraints and flows of matter and energy.

A next step is to realize that the only well-known autonomous agents, namely real cells such as yeast, bacteria, your cells and mine, do actually carry out linked processes in which spontaneous and nonspontaneous processes are coupled to build constraints on the release of energy. The energy, once released, constitutes work that propagates to carry out more work, building more constraints on the release of energy, which when released constitutes work that propagates further.

Figure 4.3 is a schematic representation of a cell. The figure shows a typical bilipid membrane, small organic molecules of different species, A, B, C, D, E, F, G,

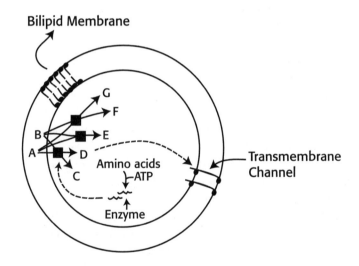

FIGURE 4.3 Cells carry out propagating work and constraint construction (see text). Cell is bounded by a bilipid membrane whose synthesis required work. Partition of small molecules, A and B, into the membrane alters their vibrational, rotational, and translational motions, hence, alters the potential barriers for three reactions, to products C and D, or E, or F and G. An enzyme, synthesized at energetic cost, binds the transition state of the A + B -> C + D reaction, lowering that potential energy barrier and allowing the constrained release of energy. That release of energy propagates to do work as energized small molecule, D, diffuses to the transmembrane channel, binds to it, does work on the channel to alter its configuration to allow ions to enter the cell.

a transmembrane channel, and so forth. Now, in fact, your cell typically does thermodynamic work to build up lipids from smaller molecular species. Typically, the energy is supplied by breakdown of ATP to ADP or similar exergonic reactions in metabolism. But lipids have the capacity to fall to a low energy structure, which is precisely a bilipid layer. As noted in chapter 3, lipids are molecules with a hydrophobic tail and a hydrophilic head. The hydrophilic head, as the name implies, likes water. Consequently, in an aqueous environment lipids will tend to form bilipid membranes with the hydrophilic heads facing the aqueous medium and the hydrophobic tails buried next to one another, away from the water. In fact, if you take some cholesterol, or another lipid or lipidlike molecule, and dissolve it in water, bilayer membrane vesicles form spontaneously that are called liposomes. So, your cells do thermodynamic work to make lipids, which spontaneously form a low-energy structure, the membrane.

But the membrane constitutes constraints. Watch. A and B are small organic molecular species and are capable of three hypothetical reactions. A and B can undergo a two substrate–two product reaction to form C and D. A and B can ligate to form a single product, E. Or A and B can undergo a different two substrate–two product reaction to form F and G. Naturally, each of these three reaction pathways from A and B passes along its own reaction coordinates through its own different "transition state." Because each of the three transition states has a higher energy than does A and B or the products C and D or E or F and G, the transition state energy is a potential energy barrier, slowing the reaction from A and B down any of the three reaction pathways.

Let A and B dissolve in the bilipid membrane from the aqueous interior of the "cell." Once this happens, immersion of A and B in the membrane environment alters the vibrational, rotational, and translational motions, or "degrees of freedom," of A and B. But, in turn, these alterations in the motions of A and B alter the heights of the transition state energies along each of the three reaction pathways from A and B to C and D or to E or to F and G.

But the alteration in potential energy heights along the three different reaction pathways from A and B is precisely the alteration of the constraints on these reactions. The barrier heights, together with the even higher energy barriers that provide the walls of the reaction coordinates along which the reaction proceeds, constitute the constraints. So, in fact, the cell has actually done thermodynamic work to construct constraints on the release of chemical energy stored in A and B, that might be released to form C and D or E or F and G.

Moreover, the cell does thermodynamic work, utilizing ATP degradation to ADP, to link amino acids together into a protein enzyme. The enzyme diffuses to the A-and-B-laden region of the membrane and binds stereospecifically to the transition state leading from A and B to the products C and D. By binding the transition state complex of this reaction pathway, the enzyme lowers the potential barrier for the A + B ↔ C + D reaction, and the chemical energy stored in A + B is released to form C + D.

Thus the cell does work, both to construct constraints and to modify those constraints, by raising or lowering potential barriers such that chemical energy is released. More, the released energy can, and often does, propagate to do work constructing more constraints. Thus, the product D may itself diffuse to a transmembrane channel and bind to the channel, giving up some energy stored in its structure by an internal rotation to a lower energy state, and thereby both bind the channel and add energy to the channel to open the channel such that calcium ions can enter the cell. A spontaneous and a nonspontaneous process are coupled. Work propagates in cells and often does so by the construction of constraints on the release of energy, which when released constitutes work that propagates to construct more constraints on the release of energy.

Records

Let's turn to the concept of a "record." As we saw, Zurek has led us to the point, in thinking about Maxwell's demon, at which a record of measurements might be kept and used later to extract work. In the case of a nonequilibrium system, in principle, measurements of a system might pay in the sense that more work could be extracted from the system—which now becomes a provider of energy—than need be used to record and later erase the measurement.

Interestingly, the "erasure cost" suggests that autonomous agents must be finitely displaced from equilibrium to afford the finite erasure cost and still reproduce. In addition, of course, rapid reproduction requires finite displacement from equilibrium.

We have many colloquial notions of a record. I want to try a tentative technical definition: Records are correlated macroscopic states that identify sources of energy that can be tapped to extract work.

Thus, we are to think of records as recording "measurements" that identify the source(s) of energy in the measured system, which may then be tapped to do work. My example of the wind through the window in Maxwell's two-chambered gas system is a case in point. We have good grounds from Zurek's work to believe that the complexity of the record is related to the reduction in entropy of the measured system.

Notice some interesting features of records. First, a useless feature of a nonequilibrium system with respect to extraction of work may be recorded. Second, errors may be made in the record of a useful feature of a nonequilibrium system from which work can be extracted. Third, the record may go out of date, so that work can no longer be extracted by reference to the record. Fourth, the record may be erased and may be updated. All the above features arise in a coevolving microbial community. Indeed, all sorts of signaling pathways in cells record and report energy sources and coordinate cellular activities within and between cells in a community. Mutation, recombination, and selection are means to update the recording devices with respect to changing sources of energy, opportunity, and danger. Again we see that cells in a community have the embodied know-how to get on with making a living.

We are struggling with a circle of concepts involving work, constraint, constraint construction, propagating work, measurements, couplings, energy, records, matter, processes, events, information, and organization. It has been said by many that we do not understand the linking of matter, energy, and information. The circle above points at something we must trouble ourselves to understand, and I suspect that the triad of matter, energy, and information is insufficient. Rather, the missing "something" concerns organization. While we have, it seems, adequate concepts of matter, energy, entropy, and information, we lack a coherent concept of organization, its emergence, and self-constructing propagation and self-elaboration.

If we do not yet understand organization fully, we can at least think about what happens in autonomous agents such as real cells. A real cell, a real molecular autonomous agent, does in fact carry out self-reproduction. In addition, it carries out one or more real work cycles, linking spontaneous and nonspontaneous processes. It does, in fact, measure, detect, and record sources of energy and does do work to construct constraints on the release of energy, which when released in the constrained way, propagates to do more work, often constructing further constraints on the release of energy or doing work by driving further nonspontaneous processes. Cells do achieve propagating work.

The work propagating in a cell achieves a "closure" in a set of propagating work tasks such that the cell literally constructs a rough copy of itself. In a later chapter I will return to discussing "tasks," which turn out on a Darwinian analysis to be a subset of the causal consequences of the release of energy at a point and time in the system. For the moment, I want to focus on the concept of a closure in a set of propagating "work tasks."

We know what it means to cook dinner, eat dinner, and clean up afterward. A coordinated set of activities is carried out that completes the events concerning preparing, eating, and cleaning up after dinner. The notion of completing a set of tasks is not mystical. So we can straightforwardly state that a cell completes a set of propagating work tasks such that it builds a copy of itself by linking spontaneous and nonspontaneous processes in constrained ways.

Thus, a molecular autonomous agent achieves two different closures. First, it achieves a "catalytic" closure; all the reactions that must be catalyzed are catalyzed by molecular members of the system. Second, it achieves a closure in a set of propagating work tasks by which it completes the construction of a rough copy of itself. Cells achieve this work-task closure, nor is there anything nonobjective about this truth.

Notice that the closure in catalytic and work tasks cannot be defined "locally." No single reaction, no single linking of spontaneous and nonspontaneous processes typically suffices to specify the closures we are describing. These closures are typically collective properties of the entire autonomous agent in its environment. In fact, cells achieve closure in some wider range of tasks by which they propagate their organization. Thus, cells carry out measurements and record them all the time. The bacterium swimming upstream in a glucose gradient was my initial candidate example of an autonomous agent. The bacterium does so by molecular "sensors" that measure glucose, a molecular motor with a stator and a rotor that can rotate in either direction, and a flagellum that can rotate in two directions, causing "swimming" in one direction and "tumbling" in the other. The cell achieves swimming "upstream" by continuing to swim if the glucose concentration is rising and tumbling then swimming in a random direction if not.

Autonomous agents achieve catalytic and propagating work-task closures by which they build copies of themselves. The myriad sensors, receptors, ligands, enzymes, and linked reactions of metabolism are the structure and dynamic of the reproducing cellular autonomous agent that constitutes the measurement, detection, recording, and search for useful energy sources to link into its ongoing construction of itself. The propagating closure of events and organization that is a cell or colony of cells, an autonomous agent, or a collection of autonomous agents is not matter alone, energy alone, entropy alone, nor the negation of entropy, Shannon's information, alone. The propagating closure that is an autonomous agent appears to be a new physical concept that we have not known how to see before.

What we can here see is the natural embodiment of organization. We have, I suggest, no coherent concept of organization. We have tended to think that the concept of entropy, of order and disorder in statistical arrangements of states of affairs, is the proper and central concept of organization. But I claim that entropy is not yet adequate. Nowhere does entropy cover the topics we have discussed, the closure of catalysis and propagating work tasks creating the complete whole that is an autonomous agent coevolving in a biosphere. This closure of tasks, measurements, records, and linkages that propagates macroscopic work seems to constitute at least an ostensive definition, a definition by example, of "organization."

Although my discussion above about organization is still preliminary, the basic points seem correct. A coevolving mixed microbial community that existed some three billion years ago, diversifying and coevolving via Darwinian mutation, recombination, and natural selection, did, in fact, measure and detect and create an increasing variety of energy sources, did, in fact, couple those detected energy sources into work cycles and other activities, and did, in fact, build a biosphere. Self-constructing organization did and does propagate. Our globe is covered by this propagating organization—life and its consequences.

Indeed, it seems important to wonder which conditions in a nonequilibrium universe would allow such propagating organization to proliferate. A biosphere does it, of course. One can imagine a watery planet with small sail boats, sails, and tillers trimmed to tack forever on a left tack, forever circling the everywhere ocean. Here the sails and tiller match the windmill and its vane, orienting the fan to capture the transient wind and extract mechanical work. Intuitively, it seems unlikely that such a planet of nonliving complex entities could have arisen spontaneously since the big bang. Just as intuitively, all we have discussed seems sufficient for the ongoing diversification of propagating organization: the Darwinian processes of natural selection and random variation, the coevolutionary construction of vastly complex autonomous-agent cell systems that continually evolve ever-novel measurements of novel sources of energy, recordings of those energy sources, couplings to those sources, constraint construction, and the linking of exergonic and endergonic reactions that builds the diversifying biosphere.

The biosphere is the most rambunctiously complex, integrated, diversifying, milling, buzzing, busyness in the universe that we know. Perhaps there are other biospheres, and they too hum in persistent diversification. Autonomous agents appear to be a sufficient condition for application of this concept of organization, and a biosphere comprised of coevolving autonomous agents appears to be a sufficient condition for propagating self-constructing organization. It remains an open question whether other structures and processes in the universe that may not be autonomous agents—say, lifeless galaxies, stars, the giant molecular clouds in galaxies, or lifeless planets—can generate and propagate diversifying organization as radically well as do biospheres.

I close this chapter by asking whether there is a way to "mathematize" the concept of an autonomous agent and, through it, the concept of propagating organization. The answer is, perhaps, category theory. I am honored to note, in memorium, that my friend and colleague Robert Rosen first explored some of these issues and some others of those touched upon here in his book *Life Itself*.

Category theory is a branch of mathematics concerning mappings. Consider a "domain" and a "range." A mapping takes points in the domain to points in the range. The mappings might be 1:1, or 1:many, or many:1. For example, in a 1:1 mapping, each point in the domain maps to a single corresponding point in the range. The domain and range can be discrete sets or continuous.

An interesting feature of categories is that a category can have the property that the mapping from the domain to the range is specified by the category itself in a recursive way; the elements of the range determine the mapping from the domain to the range. This recursive specification comes close to an autocatalytic set. We need merely think of a set of molecular species in the domain and a set of molecular species in the range; the mapping from domain to range is just the set of reactions that transform the initial "substrate molecules" in the domain to the "product molecules" in the range. Now, an autocatalytic set has the property that certain product molecules in the range, namely the products that are also catalysts, "choose" the reactions that are catalyzed from the substrates to the products, hence, choose the specific mapping from the domain to the range. Thus, an autocatalytic set can be thought of as this kind of recursive category.

The category theory image is at least a start with respect to catalytic closure. Perhaps some enhanced category theory that includes closures of work tasks, measurements, and records, as well as catalysis, is part of what an adequate formalization of "autonomous agent" may be. It is too early to say.

On the other hand, I am not persuaded that category theory will suffice. In category theory it seems necessary to specify ahead of time all the possible domains and ranges and mappings under consideration. I will suggest in a later chapter when we consider the evolution of novelties that there is no finite prespecification for the work tasks, measurements, records, and catalytic tasks that might constitute autonomous agents. In short, I will argue that we cannot prestate the configuration space of a biosphere. Whether an incapacity to prestate the configuration space of a biosphere genuinely precludes the use of category theory to mathematize the concepts of autonomous agents and propagating organization is an open question.

We have arrived at this: An autonomous agent, or a collection of them in an environment, is a nonequilibrium system that propagates some new union of matter, energy, constraint construction, measurement, record, information, and work. It is a new organization of process and events. The collective behaviors of coevolving autonomous agents have, over the past four billion years, constructed a biosphere. If life is common, the elaboration of biospheres in the universe is rife. The propagating union of work cum record cum measurement cum constraint construction, the propagation of organization unfolding and diversifying, exhibits the very creativity of the universe. We are entitled to ask whether there may be general laws governing such nonequilibrium self-constructive processes in biospheres and the universe as a whole. I return to candidate general laws in chapters 8 and 10.

Chapter 5

A PHYSICS OF SEMANTICS?

Each of the chapters of *Investigations* broaches new territory. Each is tentative and incomplete, pointing but not fully adequate. Yet, I persistently hope, it is better to light one candle against... what? the darkness? the veil beyond which we have had no framework of questions before? There are grounds, reasoned about by the best of philosophers and scientists of the past several centuries, to doubt a physics of semantics. We are on shaky ground. Yet when the first hard frost comes, when the birches have been swung, their crimson leaves scattered carelessly, when crystals splint shallow ponds and old egrets stand dark watch the coming dawn, long-legged, knowing how winter is thin, when the first phase transition of water to ice forms slight solidity across meadow streams, small creatures of flesh and concept tiptoe gingerly to some far side where, perhaps, something new is to be found.

We have lacked a physical definition of an autonomous agent, able to manipulate the universe on its own behalf—the egret whose foreboding of winter leads to lifted wing and steady, powerful flight. The egret is as much a part of physical reality as the atom, and perhaps more than the vaunted quark. But autonomous agents, we who do daily manipulate the world on our own behalf, we to whom "intentionality" and "purpose" are so inevitably attributed by our common languages, we are, by my definition of autonomous agents, also nothing but physical systems with a peculiar organization of processes and properties. If the concept of autonomous agents were something like a useful—or more, a proper—definition of life itself, then autonomous agents span the gap from the merely physical to that new realm of the merely physical where "purpose" is ascribed by all of us to one another.

Semantics enters with purpose. For this to be true, it is not necessary that the carriers of purpose, say, the same bacterium heading upstream in the glucose gradient, be conscious.

I hope my definition of an autonomous agent is useful, an autocatalytic system carrying out a work cycle, now rather broadened by the realization that autonomous agents also do often detect and measure and record displacements of external systems from equilibrium that can be used to extract work, then do extract work, propagating work and constraint construction, from their environment.

Know-how

Bring ourselves empathetically and objectively back three billion years to the mixed microbial community flourishing right about where most of us are now, plus or minus a modicum of layers of surface crustal material.

I want to say that the autonomous agents comprising that community had, individually and collectively, the embodied know-how to get on with making a living in the natural games that constituted their world. Indeed, as I have emphasized before, a biosphere is a self-consistent coconstruction of autonomous agents, ways of making a living, and the search procedures, mutation, recombination, as well as behavioral search open to autonomous agents. Those means of making a living that were well searched out and mastered by the agents and their search procedures became the kinds of "jobs" that were widely filled, the abundant niches of the biosphere. There is in this whole self-constructing system a wider know-how, beyond the know-how of any single autonomous agent spinning eagerly in its microenvironment. Yet, clearly, the know-how is distributed. There is no autonomous agent, no one, who knows how the whole system works, any more than anyone at present knows how the global economic system works in its myriad interactions, deals, steals, hopes, and frustrations.

What in the world is "know-how"? Philosophers distinguish between "know-how" and "know that." I know how to tie my shoes and am learning how to play jazz drums. "Know that" concerns propositions, most conveniently, human propositions. I know that the moon is—they tell me—not made of green cheese. I know that the earth circles the sun, that the earth is roughly spherical, that chairs are used to sit on. "Know that" brings with it the standard and nonstandard issues of the truth or falseness of propositions as they report states of the world. Perhaps higher primates who are trained to manipulate simple symbols with apparent reference to the world also can "know that" with respect to propositions.

Unlike "know that," "know-how" does not involve propositions about the world. "Know-how" involves procedural knowledge about how to get on in the world. The cheetah streaking after the wildebeest, the athletic genius high jumping, have the know-how to do it.

Does a bacterium know how to make a living in its world? I certainly want to say yes, without attributing consciousness in any way. Watch the myriad subtle turnings on and off of genes, metabolic switching, mechanical twitching, sensing of glucose gradient, swimming and tumbling upstream to higher glucose concentrations. It knows how all right, even if it cannot talk about how it gets on with its business. But then, try to talk about tying your shoes or the skilled driving when you become aware some dozens of miles down the road that you have accomplished the tasks without paying the slightest focused attention.

Thank God for know-how. Know that is a thin veneer on a four-billion-year-old know-how skill abundant in the biosphere. But any autonomous agent proliferating alone or in a congery of other agents, it would seem, is also graced by the selfsame know-how. If we synthesize autonomous agents in the next decades and they coevolve under our rapt gaze over months or years into a modestly complex ecosystem brimming with novel life forms, they too will know how to make a living in their mutually created world plus the boundary conditions we more or less intelligently impose on them.

The know-how is, in these terms, nothing but another view of the propagating closures of catalysis, work tasks, sensing, recording, and acting that we now recognize as inherent in the *doings* of autonomous agents. The know-how is not outside that propagating organization. The know-how *is* the propagating organization.

Semantics

All of which brings us, inevitably, to the brink of semantics.

It is simple at its roots, you see. An incoming molecular species arriving in the interior of an autonomous agent really is (i) food; (ii) poison; (iii) a signal; (iv) neutral; (v) something else.

Once there is an autonomous agent, there is a semantics from its privileged point of view. The incoming molecule is "yuck" or "yum." I think the major conceptual step to yuck or yum is unavoidable once there is an autonomous agent. And I think we have roughly the Darwinian criteria in mind. If yum, then there will probably be more of this type of agent, offsprings of the first. If yuck, it is not so likely this lineage will prosper.

Once yuck and yum, we are not far from C. S. Pierce's meaning-laden semiotic triad: sign, signified, significans. Like it or not, the glucose gradient is a sign, a predictor, of "more glucose that way." Granted, the glucose is not an arbitrary symbol, any more than a cloud is an arbitrary symbol of rain. In this restricted sense, signs are causally correlated with that which is signified. By contrast, the relation between the word "chair" and that which it signifies, and on which I am now sitting, is arbitrary. But can chemical signals in bacterial and plant and human communities be arbitrary from a chemical causal point of view? If so, can "mere chemicals" be signs in the full Piercean sense?

I believe it is clear that mere chemistry in an autonomous agent can harbor symbols and signs in the full senses of the words. Consider first the famous genetic code. Triplets of nucleotides in an RNA molecule stand for specific amino acids that will end up incorporated into a protein. The detailed causal machinery involves transfer RNA molecules with their anticodon site and the distant site to which amino acids are attached, the aminoacyle transferase enzymes that charge the amino acid binding site of each transfer RNA with the proper amino acid among the twenty amino acids, the binding of the charged transfer RNA's anticodon site to the proper RNA code word triplet, the ribosome that glides between adjacent charged transfer RNA molecules and links the successive amino acids into the growing polypeptide chain that hangs free in the cytoplasm, tethered by the ribosome to the messenger RNA molecule as it is "translated."

The arbitrariness of the genetic code is exemplified by the evolution of novel transfer RNA molecules, which translate a given messenger RNA code word triplet into a different amino acid. As J. Monod properly emphasized three decades ago in a slightly different context concerning activation and inhibition of enzymes at allosteric sites on the enzyme that are distant from the catalytic site, the relation of chemical structures that achieve control of catalysis are utterly arbitrary with respect to the chemical structures that undergo the catalysis. The same is true of the transfer RNA where the anticodon site is distant from the amino acid binding site. Because of this, which amino acid is charged onto a particular transfer RNA is utterly arbitrary and controlled by the aminoacyle transferase enzyme that does the charging, plus the structure of the amino acid binding site on the transfer RNA. Both of these can be altered without altering the anticodon-codon matching mechanism. In short, chemistry allows arbitrary organizations of control relations.

It seems fully legitimate to assign the concepts of sign, signified, and significans to the genetic code. It seems legitimate to extend that notion to much of the subtle signaling, chemical and otherwise, within and between autonomous agents, as exemplified by plants that upon infestation by a particular insect secrete a secondary metabolite chemical that "warns" other members of the same species that an insect infestation is happening and to turn on defensive anti-insect secondary metabolites.

The calculus that is Claude Shannon's elegant information theory has always been about reduction of uncertainty about the statistics of the source of a set of symbols. Nowhere in the core of Shannon's work concerning the encoding and transmission of information does the meaning, or semantics, of the information enter. This is no criticism, and is widely known and appreciated. There is, however, just a hint of semantics in Shannon's view that the semantics resides in the "decoder."

I cannot buy Shannon's view unless the decoder is an autonomous agent. If not, then the decoder merely transforms a bit string sent along a communication channel into some other discrete or continuous dynamics—perhaps a set of water-filled bowls is drained by turning on a machine that opens valves between the bowls and to the outside world in particular ways. The patterns of bowl drainage upon receipt of the binary string messages sent along the communication channel constitutes the decoding.

But if the recipient is an autonomous agent such as a bacterium and the incoming molecule is a symbol-sign of a paramecium or an amoeba on the roam and the bacterium swims away and avoids becoming dinner, that sequence of events seems laden with semantics. If only the bacterium could tell us: "Did you see that truck of a paramecium coming at me? I've run into that one before! I ducked under a boulder, and he never sensed me. I made it home. Pass me some more glucose please, Martha."

I will return in the next chapter to discuss such stories, for I will say that we cannot prestate the configuration space of a biosphere and, therefore, cannot deduce that which will unfold. Thus, among other things, we must tell stories to understand the oriented actions of agents in their worlds. Do not be overly quick to accuse me of anthropomorphizing. I too know the risks, including the common claim that we can always in principle translate from "intentional talk" to the fully predictive causal account of the events in question. But patience. Not only are we unable to prestate the configuration space of a biosphere and predict what will unfold, but we also cannot even translate—in the sense of necessary and sufficient conditions—from legal talk to normal intentional talk, let alone from legal talk of Henderson found guilty of murder to a physical talk about sound-wave forms monitored and masses at space-time lines as a description.

Then let's just be naive for the moment. The semantics of the yuck or yum coming into a simple autonomous agent—say, an early bacterium—is somehow linked with the embodied know-how of that agent in making a living, or failing to make a living, in its world. The semantics of an event is some subset of the fully embroidered, context-dependent set of causal implications of the event, or signal, in question.

"For want of a nail, the shoe was lost; for want of the shoe, the horse was lost; and for want of the horse, the rider was lost," more or less, said Benjamin Franklin. The semantics of the nail is some subset of this embroidered context-dependent set of implications of the event, or signal, in question to the autonomous agents in the coevolving system.

It seems hard to ascribe purpose, in the sense of acting on its own behalf, to a stone or a chair and easy with respect to an alga. Of course, there is a sense in which my attribution of semantics to autonomous agents is purely tautological. After all, I began by stating that a bacterium swimming upstream in the glucose gradient was acting on its own behalf in an environment, defined an autonomous agent as a physical system able to act on its own behalf, then asked what a physical system must be such that it can act on its own behalf. Now, five chapters later, it is hardly an independent deduction that autonomous agents are the proposed organization of matter, energy, and organization to which purposes can be ascribed in the sense of being able to act on their own behalf. My definition is a definitional circle.

On the other hand, while the definition is circular, like F = MA and Darwin's "natural selection" and "fitness," that does not mean that the set of codefined concepts surrounding my definition of an autonomous agent as a reproducing system that does a work cycle fails to touch the real world. Stones and chairs are not, by my definition, autonomous agents. All living cells are. And the stunning fact directly before us, every day, is that autonomous agents do manipulate the world on their own behalf. Watch a pair of nesting birds build their nest.

In short, once we have autonomous agents and yuck and yum, it appears that semantics enters the universe as the agents coevolve and behave on their own behalf with one another in the unfolding of a biosphere.

Knowing

What about "knowing"?

Daniel Dennett, in his fine book *Darwin's Dangerous Idea*, advances a hierarchy of forms of "knowing," if I may use that term, that have arisen in evolution by Darwinian means. I find his hierarchy congenial and informative. Dennett envisions evolution as a sequence of kinds of construction cranes that bit by bit build up higher-order entities via variation and natural selection.

I do not disagree, although I have placed far greater emphasis on the roles of self-organization in evolution. Here, in *Investigations*, I am trying to point at the mysterious but utterly natural hopefulness in which an increasing diversity of broken symmetries in the universe creates the diversity of structures and processes that can constitute and identify ramified and ramifying sources of energy, detect those sources of energy, create devices and processes that couple to those sources of energy, and generate yet more diversity that propagates macroscopic order even further. I wonder, in short, at the naturalness and self-generaticity of Dennett's cranes building cranes building cranes in biospheres, perhaps planetary geologies, and beyond.

But back to the past. Dennett distinguishes "Darwinian creatures," "Pavlovian creatures," "Popperian creatures," and "Gregorian creatures." A simple autonomous agent, say, a bacterium, is a Darwinian creature. In its simplest version, the creature evolves by mutation, also recombination and natural selection. For the moment, no behavioral learning is to be considered. So one (or a colony or an ecosystem) of Darwinian creatures adapts more or less as Darwin told us.

At the next level up, say, aplysia, a nervous system is present, and the creature is capable of stimulus-response learning, à la Pavlov. Indeed, aplysia can learn very simple conditioned stimuli—the later analogue is the bell causing the dog to salivate in "expectation" of food.

At the next level (Dennett, perhaps properly, reserves this for us vertebrates) is the Popperian creature. Popperian creatures, in Dennett's fine phrase, have "internal models" of their world and can "run the internal model" with the clutch disengaged, rather than running the model in real time in the real world. This allows us lucky Popperian creatures to allow our "hypotheses to die in our stead." I love that image.

Beyond the Popperian is the Gregorian creature—namely, at least humans. Dennett makes the wonderful argument that we utilize our tools—literally stone knives, arrows, digging sticks, machine tools—to enlarge our shared world of facts and processes. This enlarged shared world gives us more know-how, and more know that. Cultural evolution, at some point, begins to burst out-of-bounds. Hard rock music jangles the minarets of Iran. Who knows what new cultural forms will blossom? Chinese cooking lands in Cuba, and Cuban-Chinese cuisine is invented. What's next under the sun? Who can say?

I very much like Dennett's ladder of know-how, and eventual know that. Without invoking consciousness, not because it is not worth invoking but because so little sensible has ever been said on the subject, it seems worth asking how much of this hierarchy could be realized by simple molecular systems, even without evoking nerve cells.

I would think a lot of this hierarchy could find molecular realizations. For example, bacteria and amoebae do have a kind of Pavlovian learning already, for they have receptors that accommodate to a constant level of a given signal ligand and sense instead a change from the current level. This is not yet the association of a more or less arbitrary conditioned stimulus with an unconditioned stimulus, but I can imagine chemistry to accomplish the latter. As neurons are supposed to proliferate and form novel synaptic connections that survive if used and to mediate the linkage of conditioned to unconditioned stimulus, why not envision a complex chemistry, say, very complex carbohydrate-synthesis patterns sustained by complex sets of enzymes whose activities are modulated by the different carbohydrates themselves, which is true of contemporary carbohydrate metabolism. Such a system might blindly try out variant patterns of synthesis until it could establish a self-sustaining web linking the carbohydrates, the enzymes, and certain protein receptors mediating the linkage between unconditioned and conditioned stimulus, then maintain that linkage by positive feedback loops. The image is not too far from how it is imagined that "idiotype" and "anti-idiotype" immune networks work to sustain synthesis of a set of desired antibodies against an incoming pathogen. In such networks, for which there is modestly good evidence, a given first antibody serves as an antigen that stimulates the body to produce a second antibody that binds to the unique amino acid sequences, called the "idiotype" of the first antibody. In turn, the second "anti-idiotype" antibody stimulates a third, which stimulates a fourth. But this series is likely to form feedback loops because the first and third antibody can often both bind to the same site on the second antibody, hence, the first and third antibodies are similar shapes in shape space. It is not much of a stretch to think of the immune system as a conditioned stimulus response system.

Popperian creatures? Why cannot the molecular-sensing and hypothesis-testing churning concerns of the bacterium as it senses a paramecium churn twenty cycles before kicking in the rotary motor, or not, such that the wee bacterium hides under a boulder of a grain of sand until the beast passes by? Are nerves necessary? Plants, as noted, are said to signal one another with complex secondary metabolites to characterize the particular kinds of insects infesting the glade. There are arbitrary structural relations between the metabolite and the insect, just as symbols in human language are often arbitrary with respect to the signified. Not bad for nerveless nonvertebrates.

I do get stuck at Gregorian creatures. Even here, the free and open creating of new symbol strings in a language, wherever new sentences can be created, is not that fundamentally different from the persistent open creation of new kinds of molecules in the biosphere as a whole. If the conversation we recent two-legged ones are having with respect to our digging sticks and atomic bombs is impressive, so too is the chemical conversation in any full-fledged ecosystem, where we are all instrumental in the lives of one another.

I suppose I am naively driven to consider that the biosphere, with its urgent diversity in which, emboldened by all our know-how, we do get on with a very rich conversation, may very early already have harbored all the levels of which Dennett speaks. We humans are just more gregarious with our vocal cords and e-mail, I guess. Smart place, a biosphere, lots to talk about. Four billion years of yammering. Slapstick comedy may have started a long time ago.

Ethics

And what of ethics? Does a whiff of ethical issue arise with autonomous agents? Yuck or yum from my point of view if I am an autonomous agent. There are deep reasons for caution. Hume told us long ago about the "naturalistic fallacy": One cannot deduce "ought" from "is." From the fact that mothers care for their young, we cannot deduce that they ought to do so, Hume argued. From the fact that Hitler set out to conquer Europe and more and to kill Jewry, we cannot deduce that he ought to have done so.

Indeed, Hume's injunction underlies the caution of scientists about making ethical statements. We scientists find the facts. You citizens across the globe can argue the ethics. But if Hume warns us not to deduce ought from is, where do values come from at all? Hume's injunction against deduction from is to ought nevertheless began by recognizing the legitimacy of the category "ought."

The efforts following Hume to understand the meaning of ethical assertions have been long, twisted, arduous. Following the dictates of the logical positivists of the Vienna Circle that only those statements capable of verification were meaningful, philosophers as famous as G. E. Moore came to wonder if ethical assertions were merely emotive utterances. "It is wrong to kill." Becomes, "Agggah!" Does the positivist argument seem persuasive? It has always amused me that the core injunction of the logical positivists, "only those statements that are empirically verifiable are meaningful," is *itself* not empirically verifiable. One is reminded of something about hoisting and petards.

John Rawls of Harvard has argued eloquently that our human notions of fairness derive from what we would all "contractually" agree to, were we to know before birth that we would all be born with differing abilities and endowments. Thus "equality before the law" is one contract that Rawls commends to us. "Equality before the law" is far more refined than the yuck or yum of the bacterium. The emergence of ethics in the evolution of life on this planet is a fascinating issue.

I will content myself with wondering where "value" and the rudiments of "intentionality" come from in the physical universe in the first place and leave social contracts for other efforts. Where is the place of value in a world of fact?

So, a short soliloquy. Facts are know-that statements. But know-how preceded know that. While fully aware of Hume's injunction, I think that from the autonomous agent's perspective, yuck or yum is primary, unavoidable, and of the deepest importance to that agent. I suppose we apply the Darwinian criteria. Too much yuck, this one and its progeny are gone from the future of the biosphere. Without attributing consciousness to an *E. coli*, or an autonomous agent we may create in the near future, I cannot help but feel that the rudiments of value are present once autonomous agents are around.

And again without attributing consciousness, once an autonomous agent is around is the rudiment of intentionality present? If so, another cornerstone of ethical activity has been laid. Ethical behavior requires first the logical possibility of behavior for which one is responsible. You are not responsible for acts and effects beyond your control. To act ethically, you must first be able to act at all.

But what are "acts" in the first place? Daniel Yamins is a brilliant young mathematician. Now entering Harvard, Dan spent a summer with me at the Santa Fe

Institute before he learned to drive, after an earlier summer spent in the laboratory of Jack Szostak at Harvard, where, at age fourteen, Dan was learning to evolve RNA molecules to bind arbitrary ligands. Dan and I struggled that summer to make the distinction between the "doings" of an autonomous agent and mere happenings in and around the autonomous agent. Note we say the *E. coli* is swimming upstream in the glucose gradient to get dinner. But all sorts of molecular vibrational, rotational, and translational motions are occurring. What are actions and what are mere happenings?

I do not think we were successful in drawing a clean distinction between doings and happenings with clear mathematics. But I sense that the distinction between doings and happenings, Dan's happy phrasing, is relevant for *E. coli*, tigers, us, trees, and autonomous agents in general. We will meet a similar problem in the next chapter when we attempt to distinguish between the function of a part of an organism and the other causal consequences of that part of the organism.

Strange and interesting, is it not, that these issues all seem to arise with autonomous agents but not otherwise? Granted that we here seem to confront the language game circularity alluded to earlier, yet I do truly think that the rudiments of semantics, intentionality, value, and ethics arise with autonomous agents. I do not think those rudiments suffice to jump over Hume's naturalistic fallacy. We cannot deduce ought from is in any concrete context, but I think we have the categories of ought and is in the physical universe once we have autonomous agents.

Chapter 6

EMERGENCE AND STORY:

BEYOND NEWTON,
EINSTEIN, AND BOHR?

How brazen a chapter subtitle: "Beyond Newton, Einstein, and Bohr?" Yet hints we shall find, for the science of Newton, Einstein, and Bohr remains innocent of the propagating coconstructing organization of autonomous agents, of nonequilibrium systems building a biosphere. Yet surely the biosphere is part of the universe and any general laws of the universe must necessarily encompass biospheres here, and if a general biology is necessary, elsewhere in the vastness we glimpse.

And "emergence and story"? What manner of foolishness is this? Story? Surely story is not the stuff of science. I'm not so sure. Story is the natural way we autonomous agents talk about our raw getting on with it, mucking through, making a living. If story is not the stuff of science yet is about how we get on with making our ever-changing livings, then science, not story, must change. Our making our ever-changing livings is part of the unfolding of the physical universe.

Would you rather be Einstein or Shakespeare? I'm not sure whose genius is the more awesome. I come, hesitantly, to believe we need both science and story to make sense of a universe in which we agents, part of the universe, get on with our embodied know-how, we who strut and fret our hour upon the stage. Then are heard no more? Hardly. Our successes and failures trickle, tumble, and torrentially build the future of our biosphere. We Americans, fearful of Sputnik, land men and mass on the moon. Parting, we leave mass on the moon and thereby change the orbital dynamics of the solar system and beyond. Calculate that, Newton, genius that you were. From what initial and boundary conditions would you, could you, start?

But again, we have not had, nor have we yet, a theory of the propagating coconstructing organization that is a biosphere built of autonomous agents and their shenanigans.

Oh, confusion. Perhaps a certain confusion is healthy. We have not tried to embrace all of this at once before.

Hierarchies of Autonomous Agents

For a start, there appears to be an indefinite hierarchy of autonomous agents. At least in our biosphere, there is a considerable hierarchy of autonomous agents. Consider first a single-cell organism, prokaryote, cells without nuclei such as *E. coli* in your gut. *E. coli*, my canonical autonomous agent. Next, consider a eukaryote, yeast, also in your gut, also an autonomous agent. In the passage from prokaryote to eukaryotic cell, it appears that a collection of autonomous agents came to live together permanently. Eukaryotes contain mitochondria, and plant cells contain plastids with chlorophyll. In both cases, these intracellular organelles carry their own DNA, with a slightly modified genetic code in the case of mitochondria. These facts have suggested to Lynn Margulis the now rather well-accepted hypothesis that eukaryotic cells are symbionts of two or more earlier separate autonomous agents that contributed the mitochondria, the plastids, and perhaps the nuclear structure of eukaryotes into a single novel reproducing entity, the eukaryotic cell.

The eukaryotic cell, then, is a well-behaved society of autonomous agents that are now symbiotic, hence, the eukaryotic cell is a higher-order autonomous agent, comprised of lower-order autonomous agents.

But life has burgeoned beyond single-celled creatures. The sea is filled with eukaryotic colonial organisms—multicellular, usually capable of sexual reproduction but also capable of asexual reproduction by budding clumps of cells that reform the various organs, feeding tubes, mouths, stinging cells, musculature, nerve system. Blessed proliferating profusion of ways of being.

And of course, since about 1.6 billion years ago, and surely since the Ediacrin 600 million years ago, and the Cambrian period 540 million years ago, there have come to exist us multicellular sexually reproducing juggernauts.

Tyrannosaurus rex really was a juggernaut of an autonomous agent. A blue whale isn't so trifling either. Neither is the wide-flung stand of aspen astride the hillsides above Santa Fe, largest single stand of aspen in the United States, presumably all or most of which is a linked set of trees sprouting from the spreading roots of some initial individual. Julius Rebek, a chemist now at the Scripps Institute, is fond of saying that the biggest molecule he knows of is Number 7 Illinois coal, a massive hunk of coal several miles long and wide and hundreds of feet deep. Maybe Number 7 Illinois coal is from a single, clonally linked stand of aspen cousins.

So we confront a hierarchy of autonomous agents. If our definition of an autonomous agent should include that it be "an individual" capable of reproducing, then maybe whales are about as big as such agents get in the current biosphere. That's a long way up in mass and molecules from the minimal autonomous agent I sketched in chapter 3.

How far can the hierarchy go? Who knows.

Perhaps the simplest step in this hierarchy would be the hypothetical, and now almost experimentally realized, hypercycle invented by Manfred Eigen and Peter Schuster. The hypothetical hypercycle consists of a set of replicating RNA sequences, say A, B, C, and D, each of which is actually a plus and minus strand that are template complements. But in addition to A, B,C, and D each replicating individually, A helps B replicate, B helps C replicate, C helps D replicate, and D helps A replicate. Thinking of each plus/minus RNA strand pair as a replicating cycle of two template strands, these replicating cycles are linked in the ABCD hypercycle.

In fact, the hypercycle is a small society of molecular replicators that help one another replicate, hence, it is a higher-order molecular replicating system. Reza Ghadiri has nearly created a peptide hypercycle and probably soon will achieve a real experimental example. While the hypercycle of Eigen and Schuster does not yet fulfill my definition of an autonomous agent because no work cycles are done, nevertheless we have no trouble imagining a hypercycle of autonomous agents. Indeed, presumably, the eukaryotic cell is more or less just such a hypercycle.

We have beginning mathematical models that reveal something about this hierarchical organization—although the best current models are curiously limited despite their brilliance.

Walter Fontana is a theoretical chemist trained by Peter Schuster in Vienna. Walter came to the Santa Fe Institute and made a major intellectual step called "Alchemy." In chapter 2, I described the emergence of autocatalytic sets of molecular species in a chemical reaction graph. By rather independent intellectual routes that began with physicist John McCaskill's efforts to create a computer soup of Turing machines that "operated" on one another, Walter invented "algorithmic chemistry." Naturally, and most naturally in Santa Fe, where one can be healed by means known nowhere else in the universe, Walter nicknamed algorithmic chemistry "Alchemy." Unlike the alchemy of Newton's time, Walter's works.

Here is alchemy: Walter borrowed a computer language known as "lisp." Lisp expressions can operate on one another. So expression 1 encounters expression 2. At random, it is decided if 1 will operate on 2 or 2 will operate on 1. Whichever way, after the operation, the lisp expression that was operated on typically is transformed into a new lisp expression.

You see the analogy to chemistry. The transformation of the lisp expression to a new lisp expression is rather like a chemical reaction. The operating lisp expression that does the transformation is rather like an enzyme catalyzing a reaction.

So Walter let loose a pot full of 10,000 lisp expressions in a computer. These merrily bumped into one another creating new lisp expressions. Walter, as a good theoretical chemist from the Eigen-Schuster tradition, imagined his algorithmic chemistry in a "chemostat" that would hold the total number of lisp expressions at a constant 10,000. So if extra lisp expressions drove the total above 10,000, Walter randomly chose enough lisp expressions to eliminate from the pot to keep the total at 10,000. This pruning back to a total of 10,000 provides a selection pressure for lisp expressions that are formed more often than average.

Walter let loose the floodgates of his alchemical world. A torrent of ever new lisp expressions, then, stunningly, a few, then more often, one sees an enlarging population of a subset of already-seen lisp expressions.

What had Walter found? The first thing he found were "copiers," that is, lisp expressions that could copy any lisp expression, including themselves. Once such replicators emerged, they took over Walter's steaming pot of lisp expressions. Indeed, such a lisp expression is rather like Jack Szostak's hoped-for RNA polymerase, able to copy any RNA sequence, including itself. Walter called such copiers "type-1" organizations.

Could other self-reproducing organizations emerge? Walter "cheated" and simply disallowed copier replicators to occur. Bereft of copiers, Walter's soup ripped forward again. Again novel lisp expressions came forth in profusion. Again, after a while, a recurrent set of lisp expressions emerged.

What had Walter found? He found collectively autocatalytic sets of lisp expressions, essentially identical to my collectively autocatalytic sets of polymers. Walter called these "type-2" organizations. In each such collectively autocatalytic lisp set, each expression is formed by the action of some lisp expression on some lisp expression in the set. The set as a whole is collectively reproducing.

But could one find a hierarchy beyond type-2 organizations? Further research has gotten as far as a kind of type-3 organization, which consists of two or more type-2 organizations that jointly coexist and create a kind of mutual glue of lisp expressions. The glue would not be formed by either type-2 organization alone but is the conjoint construction of the plurality of type-2 organizations.

So a modest hierarchy of algorithmic chemical systems has been found. The type-3 organizations seem analogical to eukaryotic cells that harbor different replicators, mitochondria, plasmids, nuclei, in a common mutual glue of cytoplasm and shared processes.

Curiously, no higher-order organization has yet been seen. It is deeply interesting to me that no one knows why. What is limiting the persistent emergence of novel reproducing algorithmic systems?

Indeed, there are now a modest diversity of algorithmic models, such as Tom Ray's "Tierra." In Tierra, computer strings live in the memory core, reproduce, and fight one another for space in the core. A "reaper" kills random critters. Evolution to form a variety of parasites and hyperparasites occurs, including some slightly hierarchical agents. Interestingly, again the total diversity of types of critters is limited.

As John McCaskill, another theoretical physicist-chemist in the Eigen-Schuster group, points out, no one has succeeded so far in creating an algorithmic system of reproducing entities that generates impressive hierarchical agents or persistent, increasingly complex organization. Again, no one knows why.

It is possible that the constraint to algorithmic critters may be the problem. Indeed, I will suggest that the biosphere is richer than that which can, in the normal senses I know, be called algorithmic. That which is algorithmic is effectively constructable by a formal procedure that begins with definable input "data" and is operated upon by a "program" in the Turing or von Neumann sense. But I will argue that we cannot prestate some biological analogue of the input data, nor is there some biological analogue of the program governing the unfolding of a biosphere. I will argue that the configuration space of a biosphere cannot be finitely prestated, that persistent novelty occurs in the biosphere and universe as a whole. And I will opine that if we cannot finitely prestate the configuration space of a biosphere, then something is odd with how we have been taught to do our science, for in Newtonian physics, Einstein's physics, and Bohr's physics, one can finitely prestate the configuration space in question. In chapter 10, borrowing on joint work with quantum gravity scholar and friend, Lee Smolin, I will suggest that if we cannot prestate the configuration space of a universe then "time" is real and necessary, and that the way a universe constructs itself may have analogies to the way a biosphere constructs itself.

Remember, a propagating organization that builds itself and persistently ramifies in a nonequilibrium setting is not yet a concept that we understand. We have matter, energy, and entropy, but no clear notion of propagating organization in the sense we here struggle to articulate. And because the way a biosphere gets on with constructing itself may not be algorithmic, it may be that story is part of how we must, in fact, make sense of the persistent emergence of novelty in the biosphere.

Well, that's a mouthful. We'll have to struggle below to see if it makes sense and might be correct.

I want to return to Walter Fontana's algorithmic chemistry to note an interesting feature. Let's define higher-order machines in Walter's chemistry. We might think of "bundles" of lisp expressions, where an "input bundle" of lisp expressions is fed into an "assembly-line bundle" of lisp expressions to yield an "output bundle." Nothing prevents our consideration of such bundles. Given a set of possible initial lisp expressions, say, N different expressions, the different subsets, or bundles, that are possible are just two raised to the Nth power. This 2^N set is called the "power set" of the N symbol strings. A bundle acting on a bundle may produce a bundle. In general, this is just a mapping on the power set in which "machine bundles" act on input bundles to yield output bundles; that is, the set of possible input bundles, machines, and output bundles is the set of possible mappings of the power set into itself.

Well, obviously one could get bundles of lisp expressions—complex assembly lines of lisp expressions acting as machines on complex ordered sets of bundled input lisp expressions to yield ordered sets of bundled output lisp expressions. Here the "ordering" of the lisp expressions would define the assembly-line sequence of operations of the machine lisp bundle, and the ordering of lisp expressions in the input bundle would define the order in which the machine acted on the set of lisp expressions in the input bundle. And just as obviously, if Walter got type-1 and type-2 autocatalytic sets of simple lisp expressions, one could get type-1 and type 2-autocatalytic sets of machine lisp bundles operating on one another. And if sets of machines could be ordered into "units" to act on sets of input bundles to yield sets of output bundles, still higher-order autocatalytic sets of type-1 and type-2 should emerge.

Why didn't Walter find these higher-order entities? I suspect part of the answer is because nothing in his algorithmic chemistry abets the ordering of lisp expressions into ordered sets treated as units and machines by one another. The *collective properties* of ordered sets of lisp expressions are not recognized and acted upon as collective objects by Walter's soup of lisp expressions.

But such limitations seem not to hinder the biosphere. We do witness the emergence of molecular assembly lines and molecular assemblages whose collective properties are recognized and acted upon by natural selection. The transcription and translation of the DNA code to messenger RNA and protein is one example. But there are others. Many enzymes form ordered arrays of multimolecular complexes in which a substrate is progressively passed from one to another active site along an analogy to an assembly line. These higher-order complexes of molecular devices arise because natural selection is able to act upon the collective properties of such molecular aggregates when those collective properties augment adaptive fitness. In effect, natural selection recognizes the life context in which the collective behaviors of these higher-order molecular structures are advantageous.

It is important to stress an obvious feature of the biosphere, in contrast to the algorithmic computational systems of Fontana, Ray, and others. The algorithmic systems manipulate symbols according to discrete, well-defined transformation rules, a kind of algebra-mapping symbols or symbol strings into symbol strings.

In contrast, the biosphere is built up of the doings, the embodied know-how, carryings on of autonomous agents—real physical molecular systems grafting a flow of matter and energy, constraint construction, and organization into their persistent coevolution. As we will see below—where I give grounds to think that we cannot prestate the configuration space of a biosphere, hence cannot prestate the adaptations that may come to exist in an evolving biosphere—real macroscopic physical systems that are autonomous agents may not be constrained in what they can produce, as is a formal mathematical algebraic symbol system. If so, then the "algorithmic freedom" of a biosphere is deeply important, for the science of Newton, Einstein, and Bohr all suppose prediction by algorithmic calculation.

Indeed, I suspect that the persistent innovations in a biosphere stem in no small measure from the fact that while we cannot prestate the configuration space of a biosphere, the categories relevant to its unfolding novel functionalities, the biosphere is not hampered by our failure at categorization. Unlike the well-defined and formal transformation rules of an algebra or a calculational process such as lisp, the transformation rules of the biosphere enlarge and change in ways that cannot be prespecified. As concrete examples, consider the evolution of the genetic code and consider the structure of eukaryotic chromosomes, whose complex coordinated behaviors underlie both normal mitotic cell division and the astonishing sequence of meiotic reduction cell divisions in which maternal and paternal homologue chromosomes synapse, undergo recombination, and separate such that the final sperm or egg cell receives, at random, only one homologue of each parental chromosome. The emergence of these complex macromolecular systems has altered the way evolution itself unfolds.

Richard Palmer, a physicist at Duke University and the Santa Fe Institute, has commented to me that physics is used to distinguishing the initial and boundary conditions from the "laws." But in the evolution of a biosphere, the emergence of systems such as the genetic code and meiosis seems rather like the emergence of new laws. This has led Palmer to wonder whether the distinction between initial and boundary conditions and laws is really as clean as it appears in, say, Newtonian physics.

The Furniture of the Universe

All of this has, somehow, to do with the question of "the furniture of the universe" and with the troublesome questions of "emergence" and "reductionism." These are contentious issues. Philosophers and others have struggled with these issues for years. I state merely some of their outlines.

There is a strong form of reductionism, the "x is 'nothing but' y" version. We met an example of this form of reductionism earlier in the efforts to make good on sense data and logical atomism, where the hope was to build up an epistemology based on the least questionable propositions, namely reports of sense data, "I seem to hear a middle C note now," "This seems to feel like a hard, flat surface." Then the statement, "A Windsor rocker is in the living room," is nothing but a finitely prespecified list of statements about sense data.

On this definition of reductionism, a higher-level concept is reduced to a lower-level language if the truth of a defined set of statements in the lower language is both necessary and sufficient for the truth of a statement in the higher language. If this could be done, then the higher-level statement is nothing but a shorthand for the list of necessary and sufficient statements in the lower, reducing, language. Thus, the truth of the statement, "There is a Windsor chair in my office," would be reduced to the truth of some specifiable set of statements about sense data. As we saw, this effort failed in the case of statements about physical objects and sense data. While it is relatively easy to find sufficient sense data statements, it appears to be impossible to finitely specify a set of necessary and sufficient sense data statements whose truth would be interchangeable with true statements about the Windsor chair in my office.

Wittgenstein wrote persuasively that the same systematic difficulty was lodged in attempts to reduce one language game to another, for example, from a description of a legal event to a description in terms of mere human actions to a description in terms of physical events. As we noted earlier, legal descriptions involve a web of concepts concerning guilt, innocence, responsibility, evidence, admissible procedure that are absent from a description of human actions outside of the legal framework. And descriptions of human actions—and, a fortiori, descriptions of the doings of autonomous agents, even bacteria acting on their own behalf to get dinner—seem to involve a different language game than mere descriptions in terms of physical events.

Among other critical features of the action-and-doing language game is that, compared to a hypothetical "complete" physical description, the action-and-doing description picks out the relevant features with respect to the goals of the autonomous agent. Interestingly, once we are at Dennett's level of Popperian creatures, which can have internal models of, and plans for, the future and can have their models die in their stead, we seem to have arrived at a level of organization in which action-and-goal talk becomes essential. Is there a finitely prestatable set of statements about physical events that is jointly necessary and sufficient for the truth of the statement, "The cheetah is hunting the gazelle"? Like other efforts, we will find sufficient conditions but be hard pressed to find jointly necessary and sufficient conditions.

The dictionary hints at where our reductionism ideas may go wrong: Every word in the dictionary is defined in terms of other words. How could it be otherwise? Concepts are defined in webs, somehow tacked onto the real world by ostensive definitions—definitions given by pointing to examples. We carve up the world in a variety of ways, Wittgenstein's language games, that appear not to be reducible to one another in the strict sense of necessary and sufficient conditions. And this, in turn, underlies the question whether legal systems and human actions are parts of the furniture of the universe, somehow above or in addition to the locations and motions of atoms and fields.

Even at the level of basic physical theory, the same issues arise. For example, classical thermodynamics is a well-defined science in its own right. Ludwig Boltzmann, Willard Gibbs, and others struggled to invent statistical mechanics, based on Newtonian laws operating on a set of idealized particles in the $6N$-dimensional phase space we have discussed. It is generally seen as a triumph that the classical thermodynamic concepts of temperature, pressure, and entropy were reduced to statistical features of idealized sets of gas particles: temperature becoming the average kinetic energy of the particles, pressure the momentum transferred to the walls of the vessel, and entropy a measure of the number of microstates per macrostate.

But before complete triumph is declared, we should ask whether statistical mechanics constitutes a set of necessary and sufficient statements with respect to classical thermodynamics. The answer appears to be no. While statistical mechanics based on Newtonian forces yields a set of sufficient conditions, that statistical mechanics is not jointly necessary and sufficient. David Gross and other physicist colleagues have confirmed to me that one could construct different, consistent statistical mechanics based on particles following non-Newtonian laws, all of which would be interpretable as reductions of classical thermodynamics. So, even at the heart, where reduction is supposed to have taken place, there seems to be no finitely prestateable set of necessary and sufficient conditions on a lower level over a set of possible statistical mechanics that would jointly suffice for a reduction of classical thermodynamics.

The same problem arises with Darwinian theory. Darwin tells us that evolution occurs by reproduction with heritable variation and natural selection. Our biology, based on DNA and RNA and proteins, is an instantiation, a sufficient condition for Darwinian evolution. Could we now state all possible physical systems that might be capable of replication, heritable variation, and natural selection? I think not.

There is a weaker sense of reductionism, namely the casting of an account of a higher-level object, concept, or phenomenon in terms of a sufficient, but not necessary and sufficient, set of conditions at a lower level. In this sense, statistical mechanics surely does "account for" classical mechanics. Many argue that this weaker sense of reductionism suffices. Well, suffices for what? That is a bit harder to be clear about. Roughly, the temptation is to say, "Temperature is nothing but the average kinetic energy of the atoms in the system"; that is, we appear to reduce the ontological furniture of the universe.

I suspect that this familiar ontological move is not always warranted. Let's take some cases that Phil Anderson uses to exemplify "emergence." Gold is a yellow, malleable metal familiar to all of us. Nowhere in the quantum mechanical description of atomic gold are these macroscopic properties to be found. Moreover, there is no deductive way to arrive at these macroscopic collective properties from the underlying quantum mechanics of atoms of gold. Rather, we observe the macroscopic properties, find lawful features of those properties, then attempt to link them to sufficient conditions in our quantum mechanical description of matter.

Another class of cases Anderson refers to is broken symmetries. A familiar example would be a pole standing vertically on an horizontal slab on the earth's surface. The vertical pole is unstable in the face of gravity and will soon fall. It might fall and point in any direction. Hence, the system has the full symmetry of the plane prior to the falling of the pole. In due course, the pole does fall over and points in some specific direction. Thereby, the symmetry of the system prior to falling has been broken. We cannot deduce from the symmetry of the initial state how that symmetry will be broken.

Phil entitled his article relevant to this issue of emergence "More Is Different."

Let's try it with my definition of autonomous agents. As I have already hinted, systems capable of self-reproduction and thermodynamic work cycles are presumably not limited to our current DNA, RNA, protein-based cells. Almost certainly, pure protein and small molecule systems can be autonomous agents. Almost certainly, pure RNA and small molecule systems can be autonomous agents. But perhaps self-gravitating systems, lasing systems, and a variety of other physical systems can be autonomous agents.

Again, we can give several sufficient conditions, but apparently we cannot finitely prespecify a set of necessary and sufficient conditions that would allow us to prespecify all the possible patterns of construction, constraint, and organization of physical processes and matter flows that would constitute an autonomous agent.

On the other hand, my definition does seem to afford a "postconstruction" test. Bring us a candidate autonomous agent, and we can ask of it: Do you reproduce yourself and carry out at least one thermodynamic work cycle? The answer is either yes or no and is an objective fact about the entity in question. So one begins to see a pattern here. We may not be able to finitely prespecify all the possible systems that constitute an autonomous agent, but we can recognize one when we see it. We can give sufficient, but not necessary and sufficient conditions, for all physical realizations of an autonomous agent, and we can check any specific candidate case. Further, any specific candidate case either is, or is not, an autonomous agent. The statement, "The bacterium is an autonomous agent," is either true or false.

Based on this, I want to say that autonomous agents are parts of the ontological furniture of the universe. I also want to say, with Phil Anderson, that emergence is real and utterly nonmysterious.

On this view of emergence, the autonomous agent is more than the sum of its parts, but not in the sense that the behavior of the autonomous agent is not explicable as the total organization of the parts organized into the whole agent in its environment. Rather, an autonomous agent is more than the sum of its parts in the sense that a wide variety—indeed, an indefinite variety—of physical systems could be autonomous agents in the same sense, self-reproducing systems carrying out at least one work cycle.

Now we can turn to causality. The "nothing but" version of reduction, "an autonomous agent is nothing but...," has tended historically to see causality as running only upward, from the behaviors of atoms to their causal consequences in the behavior of larger entities such as tigers.

But I find myself troubled by this view and will provide an example of why. Millions of years ago, the last female trilobite, Tomasina, was hurrying to find a good place to lay her eggs. Suddenly, Tomasina saw a hideous starfish, named Darthvader, dead ahead. "Left or right? What shall I do?" she wondered. Tomasina jumped left. Darthvader jumped right, caught Tomasina, killed her, and devoured her and her eggs.

There are no more trilobites. Moreover, when Tomasina died, she took with her the unique proteins and small molecules that were trilobite molecular species. These kinds of molecules are gone from the biosphere. So too are descendant mutant molecules that might have arisen from further speciation from Tomasina's tribe. Furthermore, other chemical reactions, which might have been catalyzed transforming some molecular species to others that Tomasina's molecules and those of her descendants might have catalyzed, have perhaps never come to be in the biosphere that has evolved over the succeeding eons.

Now, Tomasina is lost as an entire organism acting in an environment. Yet the causal consequences of her wrong guess of direction to jump have propagated downward to lower levels of organization, namely the molecular species of which the biosphere is composed.

Downward causation is real and nonmystical. There are now more old tires along roadsides than wagon wheels. The car has replaced the Conestoga.

I distrust a reductionism that sees causality as bottom up. In what sense is Tomasina nothing but the atoms and their locations and motions in three-dimensional space of which she was comprised? The concepts of atoms in motion in three-dimensional space do not appear to entail the concepts of an autonomous agent, self-consistent constraint construction, release of energy, propagating work tasks, and the closure of catalysis, tasks, and other features that constitutes the propagating organization that is an autonomous agent or a coevolving ecology of autonomous agents. In one sense, of course, there is nothing but the atoms in motion in three-dimensional-space in Tomasina. But the historical coming into existence of life in the universe, of autonomous agents, and of the propagating organization that is Tomasina and her bioworld is nowhere accounted for by Newton's laws. What, after all, do Newton's laws of motion have to do with a sufficient account of Tomasina's jump to the left rather than the right? Is Tomasina as a whole organism part of the furniture of the universe? Yes.

Adaptations, Exaptations, and the Impossibility to Finitely Prestate the Configuration Space of a Biosphere

One contentious issue leads us to another. And now I hope to trouble you deeply.

Tomasina was but one of many lineages of creatures evolving much as Darwin taught us, by heritable variation and natural selection. Let's turn to a Darwinian account of the function of the heart. Roughly, Darwin would say, the function of the heart is to pump blood. Namely, Darwin would say, this causal consequence of the heart is the virtue for which it was, and persistently is, selected by natural selection. I tend to think Darwin's account of the heart is correct. Notice that the form of the account is ontological. Hearts came into existence, somehow, and are sustained because it is advantageous to organisms to have a means to pump blood.

But the heart has other causal consequences. For example, the heart makes heart sounds as its valves open and close. Presumably, heart sounds are not the causal consequences of the heart upon which natural selection has acted. Heart sounds are not the function of the heart.

It is precisely this point, that the function of a part of an organism is a subset of its causal consequences, to which I appealed earlier in stating that in an autonomous agent discerning the work task done by the constrained release of energy required finding the subset of causal consequences of that work task that were functionally important to the life cycle of the autonomous agent in its environment and, therefore, were presumably selected and sustained by natural selection. My point was that we cannot know the functions of parts except in the context of the whole autonomous agent in its environment.

But now we come to a more radical issue, Darwinian preadaptations, or in Stephen J. Gould's term, "exaptations." Darwin noted that in an appropriate environment a causal consequence of a part of an organism that had not been of selective significance might come to be of selective significance and hence be selected. Thereupon, that newly important causal consequence would be a new function available to the organism.

Take a fanciful case in point. The human heart not only makes heart sounds, but it also is a set of resonant chambers. Suppose you were in Los Angeles, felt something odd in your chest, and thought, "My God! An earthquake!" You did whatever the right thing was and survived, alone among millions. Your heart happened to be preadapted to pick up earthquake pretremors. Now suppose you marry and have children who inherit your fatefully preadapted heart, and suppose earthquakes arose often enough for this new capacity to offer survival advantage to you and your offspring. Soon a subspecies of *Homo sapiens* would arise with earthquake detectors in their chest. Not bad.

Now, evolution by such preadaptations, or exaptations, are not rare; they are the grist of adaptive evolution. Thus arose the lung, the ear, flight, presumably most major adaptations and presumably many or even all minor ones as well. It suffices for my purposes that many adaptations arise as Darwin's preadaptations, or Gould's exaptations.

Here now is my troublesome question. Do you think that you could state, ahead of time, all the possible causal consequences of bits and pieces of organisms that might in some odd circumstances or another turn out to be preadaptations and hence be selected and come to exist in the biosphere? Stated more starkly, do you think that you can finitely prestate all the context-dependent causal consequences of parts of all possible organisms that might be preadaptations, hence be selected and come to exist in the biosphere?

I believe, and it is a matter of central importance if I am correct, that the answer is no. I do not think it is possible to finitely prestate all the context-dependent causal consequences of parts of creatures that might turn out to be useful in some weird environment and hence be selected. I'm not yet certain how to prove that this is not possible, although I will have a try at it below.

Another way of stating what I am driving at is this: Is there a finitely prestatable set of all the possible potential biological functions? Again, I think the answer is no. Yet another way of stating this is to say that there is no finite prestatement of the configuration space of a biosphere. We cannot say ahead of time all the possible constellations of matter, energy, process, and organization that is a kind of "basis set" for a biosphere in the sense that the atomic chart of the elements is a finite basis set for all of chemistry.

It is time for a story. A particularly ugly squirrel named Gertrude was atop a tree 65,433,872 years ago. Gertrude was ugly because she had folds of skin from her forearms stretching to her hind limbs. So ugly was Gertrude that she was shunned by the other squirrels and was sadly alone atop a magnolia tree eating lunch. But just yards away, high in a pine, was Bertha, an owl. Bertha spotted Gertrude and thought, "Lunch!" Bertha flashed downward through shafts of light toward Gertrude. Gertrude looked suddenly up and was terrified. "GAAAAAAH," she cried and jumped in desperation from the top of the magnolia tree, flinging her arms and legs wide in terror.

And Gertrude flew! Yes, she flew away from the magnolia tree, eluding the bewildered Bertha. Later that month, Gertrude was married in a civil ceremony to a handsome squirrel, as she had become a heroine, was no longer shunned, and was considered a prize mate. Her odd flaps turned out to be a consequence of a simple Mendelian dominant gene, hence her kids had the same wondrous capacity to fly.

And that is how flying squirrels got their wings, more or less.

Now, after the fact, after Gertrude jumped in terror from the magnolia tree, we would all say in wonder, "Did you see what Gertrude just did?!" And we would tell the story of Gertrude. But could we have said beforehand that Gertrude's ugly skin flaps would happen to be of use that day? Perhaps, perhaps not. Could we have said it four billion years ago? Or said it today about all possible future exaptations? No. Was some known law of physics violated by Gertrude? No.

Now a story about tractors: It is said, and I choose to believe it true, that some engineers were hard at it trying to invent the tractor. "Good idea, a tractor," was the collective wisdom. Needing lots of power, the engineers began with a huge engine block and sought to mount the block on a chassis. But the engine block was so massive that it crushed chassis after chassis. The engineers were stumped. Then one day, one of the engineers said, "You know, the engine block is so rigid, we could use the engine block itself as the chassis and hang everything else off the engine block. And so that's how the tractor got its chassis. And the chassis is just another Darwinian preadaptation, or Gouldian exaptation.

A brief history of the origin of writing: In the early Near East, loans of sheep and goats were common. The borrower would give the lender a small, closed vessel of baked clay, containing a number of stones equal to the number of borrowed sheep. Upon return of the sheep, the vessel would be broken open and the stones counted to make sure as many sheep were returned as had been borrowed.

But sometimes the clay vessels were broken accidentally by the lender before the time of return of the sheep. When the vessel was broken, sometimes the stones would fall out and become lost. The lender could not be sure he had recovered all his sheep. So people started making scratch marks near the top of the vessel before they baked the clay, to denote the number of stones placed inside the closed vessel. One day it dawned on someone that with the scratch marks on the surface of the clay vessel the stones inside were not needed. They smoothed the clay and began keeping notes of loaned sheep by marks baked into the clay. Cuneiform writing began.

Do you think you could finitely prestate all the context-dependent causal consequences of human artifacts that might turn out to be useful in some odd environment or for some odd purpose? I don't think so. It is not that we cannot finitely prestate some infinite things. For example, Fourier had a wonderful idea. "Sine and cosines, you know," he muttered to himself in French. "All possible wavelengths, out to infinity, down to infinitesimal, all possible phase offsets. . . . Haha!" And Fourier proved his theorem that any wiggly line on a plane surface could be approximated to arbitrary accuracy with a weighted set of phase-offset sines and cosines drawn from the infinite basis set of all sine and cosine functions. Fourier finitely prestated an infinite basis set for all continuous differentiable wiggly lines on long blackboards.

But can we finitely prestate all possible exaptations for all possible organisms, or even the current organisms, in our biosphere? Again, while I'm still not certain how to prove my claim, I claim the answer is no. We cannot prestate the configuration space of the biosphere. But notice our failure is not hindering the biosphere from exapting all the time. Gertrude did it. And every bacterium whose molecules wiggle in a useful way that turn out to detect a source of energy or danger or opportunity in some novel fashion tends to be selected for that novel functionality. Look at the rate of emergence of bacteria resistant to our antibiotics and the myriad unexpected ways such resistance arises at the molecular level. Who could have foretold the ways?

The example of the common emergence of Darwinian preadaptations in the biosphere may point to an interesting connection with the difficulty Fontana and others have had achieving the persistent emergence of more complexity in algorithmic models such as alchemy and Tierra. Gertrude did fly, and thereby the capacity of her folds of skin to function as wings were selected. Flying squirrels came to exist in the universe. Restated, a property of Gertrude—indeed, here a collective property of her atomic constituents—made itself manifest in the real physical world in a context that lent survival advantage to Gertrude. Were we to have a formal algorithmic description of a formal simulated algorithmic Gertrude that did not have as an algorithmic consequence that her skin flaps might function as wings, then the emergence of the higher-order category of "winged squirrel" could not be derived algorithmically. Similarly, were we to have a formal description of an engine block that did not include its rigidity, we could not algorithmically derive that the engine block could be used as a chassis.

But, we might ask, could we not have a complete physical description of Gertrude or the engine block such that all possible properties might be derived from that complete physical description? The answer is almost certainly no. It is an old philosophic realization that there is no finite description of a simple physical object in its context. For example, the coffee table in my living room is made of three wooden planks, four short squat legs, runners between all pairs of legs. The middle board has a crack in it some eight inches long, a quarter of an inch wide at the end of the board, narrowing to nothing along a particular curved arc. A second crack, smaller, is six inches from the first crack. A cracker is on the table. A personal computer is on the table. The first crack is seven feet from the door. The second crack is seven feet six inches from the door. Both cracks are 256,000 miles from the moon and 4.3 light years from the nearest star. A dead grasshopper is on the table to the left of the end of the first crack and about 4.3 light years from the nearest star. A mote of dust hovers an inch above the table, two inches from a leaf that drifts down from a ficus in the living room.

You get the sense that there is no complete description of the table. Why does it matter? Because I myself made an exaptation of which I am deeply proud. You see, I was worried one day that my wife or adult son might knock my PC off the table, so I wedged the power cord into the first crack and plugged the cord into a floor socket. Thereby, my PC worked on the table, and couldn't be knocked off easily. You can understand my pride here, as with my Rube Goldberg device to water my bean field.

But even this tiny invention, this tiny exaptation, could not readily have been finitely prestated. How would one, in describing all the context-dependent features of the table, happen to list the crack and its distance to the floor socket that happen to turn out to be relevant for my brilliant solution of a sudden problem? In short, there seems to be no finitely prestatable effective procedure to list all the context-dependent features of objects and organs that might prove useful for some oddball purpose by some organism. My invention, the tractor invention, the cuneiform invention, and Gertrude's invention were all genuine novelties in the universe.

This brings us to a wondrous set of issues. You see, we have indeed been taught by our physicist friends to do science by prestating the configuration space in question. Consider our now rather tired example of statistical mechanics with Avogadro's number of gas particles in a liter container. First, note that we can finitely specify ahead of time the $6N$-dimensional configuration space of the gas, that is, all the positions and momenta of the N gas particles in three-dimensional-space inside the liter box. Then Boltzmann assumed the ergodic hypothesis about wandering all over the configuration space, did the calculations, and, lo, statistical mechanics is upon us.

Now Newtonian mechanics: Prestate the initial and boundary conditions, the particles and force laws, and with them the possible configuration space and calculate away. So too in general relativity: Given Einstein's equations, prestate the initial and boundary conditions and seek solutions. Solutions are possible universes. The set of possible solutions is the configuration space allowed by general relativity. And in quantum mechanics, one talks of specifying the classical conditions of the experiment, and thereby the configuration space of the quantum system, preparing an initial state, and using Schrödinger's equation to propagate amplitudes for the entire future evolution in the configuration space for all conceivable observables. Again in quantum mechanics, in any specific context the configuration space is to be finitely prestatable, then we follow the deterministic time evolution of the Schrödinger equation in configuration space, square the resulting amplitudes to predict the probabilities of measurements, and then carry out macroscopic measurements. We know the configuration space ahead of time.

But what if we cannot prestate the configuration space of a biosphere? In that case, the way Newton taught us to do science is not the whole story. We cannot calculate as he did. And, in fact, biologists do not often do science as Newton taught. We carry on an odd mixture of historical analysis of the actual branching pathways of evolution; a dollop of theory about evolutionary landscapes, molecular evolution and coevolution, and ecosystems; and a lot of detailed experimental work to understand how actual creatures develop, how their life cycles unfold, how they assemble into ecosystems, and so forth.

And biologists tell stories. If I am right, if the biosphere is getting on with it, muddling along, exapting, creating, and destroying ways of making a living, then there is a central need to tell stories. If we cannot have all the categories that may be of relevance finitely prestated ahead of time, how else should we talk about the emergence in the biosphere or in our history—a piece of the biosphere—of new relevant categories, new functionalities, new ways of making a living? These are the doings of autonomous agents. Stories not only are relevant, they are how we tell ourselves what happened and its significance—its semantic import.

In short, we do not deduce our lives; we live them. Stories are our mode of making sense of the context-dependent actions of us as autonomous agents. And metaphor? If we cannot deduce it all, if the biosphere's ramblings are richer than the algorithmic, then metaphor must be part of our cognitive capacity to guide action in the absence of deduction.

Indeed, in biology itself the "narrative stance" is gaining in popularity. "Did you see what Gertrude pulled off? That's how flying squirrels evolved! Dominant Mendelian gene, you see, easily selected once the right environmental conditions arose." The propagating exapting biosphere is getting on with it, and it appears that we crucially need stories to do some of the telling of that getting on with it.

How odd. C. P. Snow wrote of the two cultures, science and the humanities, never to mix. Our inability to prestate the configuration space of a biosphere foretells a deepening of science, a search for story and historical contingency, yet a place for natural laws.

Forever Creative

In this chapter I have been trying to say, argue, articulate the possibility that a biosphere is profoundly generative—somehow fundamentally always creative. The cornerstone of this dawning near conviction lies in the belief I now hold with some confidence that we cannot finitely prestate the configuration space of a biosphere. New variables—the genetic code, recombination, Gertrude's wings, writing, the tractor—persistently emerge. New language games and living games emerge. What is the status of my claim that we cannot finitely prestate the configuration space of a biosphere? I do think my claim is true. But why? I am not sure. It is wise to explore some possible reasons.

A first possibility is that the biosphere, like a complex algorithm, unfolds in ways that cannot be foretold. Recall that for many algorithms the behavior of the algorithm cannot be prestated in any form more compressed than simply watching the program unfold. The famous "halting problem" is the classic example. For many algorithms that are to compute an answer, then halt, we cannot say ahead of time whether the computer will halt in finite time.

I do not think the biosphere is akin to this difficulty with many algorithms. If we consider such algorithms, the building blocks of the algorithms—for example, the binary symbols 1 and 0; the operations of addition, subtraction, multiplication, division, exponentiation, and root taking; and control operations such as, "If such and such, then do so and so, otherwise do this," and "Do loops"—are well-stated, crisp, mathematical primitives. Our uncertainty about the unfolding of an algorithm does not lie in uncertainty about the primitives, but about the consequences of the arrangements of these agreed upon primitives in a given computer code. For example, will the algorithm based on those primitives halt or not halt in finite time?

But among the exaptations in a biosphere are those that appear to alter the primitive objects and control operations. Thus, the evolution of chromosomes that replicate and partition to daughter cells, the evolution of the genetic code, and the evolution of controlled recombination all seem to be the evolution of the generative machinery of evolution itself. Insofar as this is true, our incapacity to prestate the configuration space of the biosphere is not a failure to prestate the consequences of the primitives, it appears to be a failure to prestate the primitives themselves.

Let's consider the possibility that the incapacity to finitely prestate the configuration space of a biosphere is related to Godel's theorem. Godel demonstrated that for axiomatic systems as rich or richer than arithmetic, given a set of axioms, there were always statements that were true but not formally derivable from the axioms. In addition, Godel showed that it was always possible to enrich the axiom set, and from that enriched axiom set, it would be possible to prove the formally true but unprovable statements in the formal system. On the other hand, he also showed that the new enriched axiom system would itself have still further formally true but unprovable statements.

I am not persuaded that the uncertainty about the configuration space of a biosphere is analogous to true but formally undecidable statements in a formal system. I base this upon an analogy between formal proof and causal consequences. The same parallel was pointed out by Robert Rosen in his book *Life Itself*. If we are to represent causal consequences by a formal system, then the concept of a proof derived by formal procedures from axioms—or more generally, the concept of a trajectory in a state space, where successive states along a trajectory are derived by a formal procedure such as integration of the differential equations representing the system—is the natural way to represent causal consequence. If this parallelism is taken seriously, then statements in a formal language that are true but unprovable in that formal language can have no causal pathway—that is, proof—from the axioms to the desired consequence. But this analogy seems to fail with respect to the evolution of the biosphere. There is a perfectly fine causal account of Gertrude and her maiden flight. We can reconstruct that account after the fact, even if we could not have predicted it. Thus, it does not seem that our difficulty in prestating all exaptations is the same as the mathematical fact of formally undecidable statements in an axiom system.

On the other hand, there may be a parallel between the exaptations of which we have spoken and Godel's theorem and the *augmentation* of the axiom set such that formerly unprovable statements become provable. That is, the emergence of novel exaptations in evolution do seem rather like the emergence of novel primitive objects and primitive control operations—hence, novel axioms. In the examples above, the emergence of the genetic code and the emergence of chromosomes that duplicate and partition daughter chromosomes into two daughter cells, the evolution of controlled recombination, seem to become instantiated as "biological laws," even though they are entirely historically contingent. Changing the biological laws in evolution seems rather like the generation of a novel axiom from which new consequences can be derived.

On this interpretation, my claim that we cannot finitely prestate the configuration space of a biosphere becomes the claim that the biosphere keeps generating new "causal axioms" from which it generates novel forms. Then just as we do not know where the new axioms of a formal system come from, save as the free invention of the logician involved, so it would seem that we cannot prestate the new generative exaptations that allow evolution to drive in new directions. I am not entirely persuaded by this analogy to finding ever new axioms in Godel's theorem, but it does have some coherence.

Indeed, the failure to be able to prestate the configuration space of the biosphere may be yet deeper. I will take a stab at a proof. I begin by vitiating my assumption that one cannot prestate the configuration space of a biosphere, then try to show that the implications are that the number of potentially relevant properties is vastly hyperastronomical and that there is no way in the lifetime of the universe for any knower within the universe to enumerate, let alone work with, all the possible properties or categories and their causal consequences.

Let's restrict attention to a model of a molecule, a square 10 x 10 array of magnetic dipoles, called spins, that can point only up or down. So our little system has 100 spins. Thus, I begin by vitiating my assertion and telling us what the configuration space is: it has something to do with the spin configurations. Well, how many different spin configurations are there? Two raised to the 100th power. Now what might we want to call a "category" or a "property"? A sensible thought is that a category or class is some collection of the possible spin configurations, say, the configuration with all spins up plus all the configurations with no more than two spins down. How many such possible classes are there? The answer is the power set, 2 raised to the 2 raised to the 100.

This is a gargantuan number. It corresponds roughly to 10 raised to the 10 raised to the 29th. That is, the number of possible static categories of our tiny 100-spin system is about 10 raised to the power written with 1 with 29 zeros after it. By comparison, the estimated number of particles in the known universe is about 10 raised to the 80th power.

But we have so far considered only static categories—possible subsets of states of the 100-spin system. Suppose the spins can flip. Then the spin system can pass from one configuration to another. Suppose that the motions of the spin system are confined to closed cycles in the space of configurations, that is, simple orbits. Each of our 10 to the 10 to the 29 static categories is a set of one or more spin configurations. The number of possible orbits through that set is the factorial of the number of members of the set. The number of orbits among configurations that constitute the power set of classes of the 100 spins is vastly larger than the number of static categories.

On the other hand, arbitrary motion among spin states may be unreasonable. Instead, the flow among spin configurations may be constrained by the energy couplings among the spins. Physicists will properly talk of the "Hamiltonian function" that gives the energy of each spin configuration. Grant such a Hamiltonian, say a spin-glass Hamiltonian, where a spin glass is a disordered magnetic material. Then the system at finite temperature may wander through the square root of its 2^{100} states, hence about 10^{15} states in some complex patterns.

Now consider two such molecular systems, each merely 10 x 10 spins, each governed by a spin-glass Hamiltonian function, and let the two molecular systems interact in an aqueous medium. As the two touch one another and jiggle near one another, the coupled system performs some very complex dance of spin motions. In general, the equations describing that motion cannot be solved analytically but would have to be solved by numerical simulation. That is, there is no short description of the behavior of the system of equations, they must instead be solved by an algorithmic system that follows the trajectory in the state space of the system by tiny incremental steps. It becomes easy to conjure multimolecular systems, indeed autonomous agents are examples, in which even if the Hamiltonian for each single system and all the coupled systems were known, it would not be possible to compute the detailed dynamics of the coupled spin system in the lifetime of the universe.

But it is just such detailed wiggling by the coupled system that allows discovery of the preadaptation that a particular wiggling of one molecule senses a subset of states of another molecule and is useful for some survival purpose. The behaviors of the collective set of molecules among the coevolving autonomous agents stumble upon, then reinforce by heritable variation, the odd molecular motions that capture photons, that sense energy sources, that are the fine-grained molecular exaptations that are the daily stuff of evolution.

We cannot compute it. There is a sense in which the computations are transfinite—not infinite, but so vastly large that they cannot be carried out by any computational system in the universe. Indeed, one can consider the known radius of the universe at the Planck length and Planck timescale, and can imagine all the events that have happened within any causally connected light cone that might, therefore, carry out a computation. While vast, there are combinatorial problems that are still vaster. Presumably, no physical process in the unfolding universe could have foreknowledge of all features of such problems. Nor would there be an effective procedure to prepare the cosmic computer in a proper initial state, read in the data, and read out its computation.

My best bet is that the incapacity to finitely prestate the configuration space of a biosphere is deeply related to the incapacity to enumerate and predict all the possible detailed dynamics of coupled molecular systems by any computational system in the universe. In turn, this incapacity is, I suspect, deeply related to the gargantuan nonergodicity of the historical universe that I discuss in the next chapter. And as we shall see in detail, the exaptations of Gertrude and others leaves macroscopic living footprints, propagating frozen accidents, on the history of the universe. The universe in its persistent becoming is richer than all our dreamings.

Chapter 7

THE NONERGODIC UNIVERSE:

THE POSSIBILITY OF NEW LAWS

F ROM A BIOSPHERE and its mysteries, this investigation now steps gingerly toward the cosmos. *Caveat lector:* I am not a physicist.

Our now familiar liter of gas particles at room temperature comes to equilibrium rapidly, certainly on the order of hours or days. "Equilibrium" means, roughly, that the macroscopic properties of the system, such as temperature and pressure, have stopped changing, except for small "square root N" fluctuations away from equilibrium that soon dissipate back toward equilibrium. As we have noted, thanks to the ergodic hypothesis, the gas system ultimately visits each macrostate at a number of times proportional to the number of microstates in that macrostate. For equilibrium with respect to all macroscopic properties to have been attained, it is not necessary that all microstates have been sampled, of course, but that the statistical distribution of microstates approaches the equilibrium distribution.

Physicist Richard Feynman noted that equilibrium is when "all the fast things have happened, and the slow ones have not." His dictum suggests, accurately, that the notion of equilibrium is not quite so self-evident.

The aim of the current chapter is to explore the profound failure, on the scale of a sufficiently large closed thermodynamic system and, a fortiori, the open system of the biosphere, to come close to equilibrium on the vastly long time scales with respect to the lifetime of the universe. The main facts are known to physicists, of course. The universe is vastly nonergodic above modest levels of molecular complexity, let alone with respect to gross motions of parts of the universe with respect to one another.

Given that the universe is actually nonergodic, nonrepeating, and in macroscopically important ways, over a time scale vastly longer than the lifetime of the universe, we are entitled to broach the question of whether there might be general laws governing some or all aspects of this nonergodic behavior. No one knows, but I will raise a possibility that has a chance to be true for a biosphere. There is little harm in wondering if it might hint at a fourth law of thermodynamics for self-constructing systems such as biospheres.

The Actual and the Adjacent Possible

I now want to reintroduce a central concept of alarming simplicity. Consider all the kinds of organic molecules on, within, or in the vicinity of the Earth, say, out to twice the radius of the moon. Call that set of organic molecules the "actual."

Now recall the concept of a reaction graph, a bipartite graph with nodes representing chemical species and lines called hyperedges leading from each set of substrates to a box and from the box to the product species of that particular reaction. Recall that we utilized arrows on the lines to distinguish substrates from products, but that the direction of flow of the reaction depended upon the displacement of the substrates and products of that reaction from the equilibrium concentrations for that reaction. That equilibrium ratio corresponds to the concentrations of substrates versus products, where the net rate of production of products from substrates equals the net rate of production of substrates from products.

The reaction graph is just the set of all the molecular species and all the hyperedges representing all the reactions among the species. Thus, consider the reaction graph among the molecular species in the actual, where at present presumably hundreds of trillions of molecular species exist.

Now consider the adjacent possible of the reaction graph of the actual. The adjacent possible consists of all those molecular species that are not members of the actual, but are *one reaction step away from the actual*. That is, the adjacent possible comprises just those molecular species that are not present in the vicinity of the Earth out to twice the radius to the moon, but can be synthesized from the actual molecular species in a single reaction step from substrates in the actual to products in the adjacent possible.

Note that the adjacent possible is indefinitely expandable. Once members have been realized in the current adjacent possible, a new adjacent possible, accessible from the enlarged actual that includes the novel molecules from the former adjacent possible, becomes available.

Note that the biosphere has been expanding, on average, into the adjacent possible for 4.8 billion years. Presumably, when life started there was a modest variety of a few tens to a few hundreds of organic molecular species—methane, hydrogen, cyanide, the familiar list. If there are now a standing diversity of 100 million species and each had a hundred thousand genes and genes in each species were at least slightly different from genes in all other species, then, not counting molecular diversity within species, the number of genes is 10 trillion. Given RNA, protein, polysaccharides, lipids, and other organic molecular species, the diversity is likely to be hundreds of trillions or more.

Something has obviously happened in the past 4.8 billion years. The biosphere has expanded, indeed, more or less persistently exploded, into the ever-expanding adjacent possible. The secular diversity of organic molecular species has increased, on average, over the past 4.8 billion years.

It is more than slightly interesting that this fact is clearly true, that it is rarely remarked upon, and that we have no particular theory for this expansion. Indeed, I note for future reference that the standing diversity of species in the biosphere has, on average, with noticeable crashes in large extinction events, increased over the past 4.8 billion years. And among us mere humans, the diversity of ways of making a living has increased dramatically over the past 3 million years, the past hundred thousand years, and even over the past thousand years. If you wanted a rabbit for dinner thirty thousand years ago, you bloody well went out and caught a rabbit. Now most of us can go buy a rabbit dinner. Something again has happened. At the level of species and ways of making a living in the "econosphere," the actual has expanded into a persistent adjacent possible.

We are all parts of the universe. So, in our little hunk of the universe, with the sun shining beatifically upon us, rather remarkable goings on have occurred. Indeed, the biosphere may be one of the most complex things in the universe.

Now a second simple point. The molecular species of the actual exist. Those in the adjacent possible do not exist—at least within the volume of the universe we are talking about, which we can expand in a moment to be the actual molecular diversity of the entire universe, not just our tiny patch of it.

The chemical potential of a single reaction with a single set of substrates and no products is perfectly definable, both the enthalpy and entropy. But we hardly need that sophistication. The substrates are present in the actual, and the products are not present in the actual, but only in the adjacent possible. It follows that every such reaction couple is displaced from its equilibrium in the direction of an excess of substrates compared to its products. This displacement constitutes a chemical potential driving the reaction toward equilibrium. The simple conclusion is that there is a real chemical potential from the actual to the adjacent possible. Other things being equal, the total system "wants" to flow into the adjacent possible.

If there are 10 to 100 trillion organic molecules in the biosphere and each pair of organic molecules can undergo at least one two substrate–two product reaction, then the diversity of reactions is the square of the diversity of molecular species, hence about 10^{14} x $10^{14} = 10^{28}$. Some substantial fraction of these reactions flow from the actual to the adjacent possible. The total chemical potential from the actual into the adjacent possible is hard to estimate, but it is certainly not small.

The Nonergodicity of the Universe

A further point is of fundamental importance, in my view. The universe, at levels of complexity of complex organic molecules, is vastly nonergodic.

Consider the number of possible proteins of length 200. That is, we consider proteins made of the familiar 20 kinds of standard encoded amino acids and, thus, linear chains of 200 such amino acids. Since there are 20 choices at each of 200 positions, the number of possible proteins of length 200 is 20 raised to the 200th power, or approximately 10 raised to the 260th power, 10^{260}.

Now let's consider the estimated number of particles in the known universe, which is 10^{80}. Thus, the maximum number of pairwise collisions that could occur in any instant, ignoring distances between particles, is that number squared, or 10^{160}. A fast reaction occurs in a femtosecond, or one part in 10^{15} seconds. Then the number of pairwise collisions and reactions that can have occurred since the estimated time of the big bang fourteen billion years ago is 10^{160} times the number of femtoseconds since the big bang, which is about 10^{33}.

The total number of reactions on a femtosecond timescale cannot be larger than 10^{193}, a very very big number. But 10^{193} is infinitesimally small compared to the number of possible proteins of length 200, namely, 10^{260}. In short, the known universe has not had time since the big bang to create all possible proteins of length 200 once. Indeed the time required to create all possible proteins at least once is at least the ratio of possible proteins to the maximum number of reactions that can have occurred in the lifetime of the universe, or 10^{67} times the lifetime of the universe.

Let that sink in. It would take at least 10 to the 67th times the current lifetime of the universe for the universe to manage to make all possible proteins of length 200 at least once. Obviously, with respect to proteins of length 200 the universe is vastly nonergodic. It cannot have equilibrated over all these possible different molecules.

At a level of complexity above atomic nuclei, once into the realm of complex molecules, the universe will not, cannot, come to equilibrium, on vastly long timescales compared to its historical age. Indeed, the giant cold molecular clouds in galaxies, about 4 degrees absolute in temperature, are highly complex mixtures of molecular species, many carbonaceous, as well as the birthplace of stars. We will return in a moment to wonder about whether a galaxy, considered as a closed thermodynamic system, reaches equilibrium chemically.

What about the arbitrary restriction to a femtosecond? The fastest known timescale is the Planck timescale, one in 10 to the 43rd parts of a second, or 10^{-43} seconds. At the Planck timescale, therefore, the universe can have created at most 10^{221} proteins of length 200 compared to 10^{260} such proteins. It would take the known universe, chunking along on the Planck timescale, 10^{39} times its current lifetime to make all proteins of length 200.

Now many biological proteins are of length 300, or even 1000 amino acids. Hence the number of possible proteins of length 1000 does its now familiar hyper-astronomical combinatorial explosion to 20^{1000} or 10^{1300}. The universe can have managed to make 10 to the 221 of these at the Planck timescale.

Forget it. The universe is vastly nonequilibrium, vastly nonergodic at the level of complex organic molecules. A fortiori, the universe is vastly nonergodic at the level of species, languages, legal systems, and Chevrolet trucks.

It follows that, even if we consider the universe as a whole, at the levels of molecular and organizational complexity of proteins and up, the universe is kinetically trapped. It has gotten where it has gotten from wherever it started, by whatever process of flow into a persistently expanding adjacent possible, but cannot have gotten everywhere. The ergodic hypothesis fails us here on any relevant timescale.

More, the biosphere, and the universe as a whole, may well be kinetically trapped into an evermore astonishingly small region of the entire space of the possible it might have reached. Stated otherwise, the set of actual small molecules and large molecules such as proteins that do exist now is presumably an increasingly tiny subvolume of the total set that might have arisen by now in the biosphere or the universe since the big bang.

This nonergodicity is puzzling. Just what, for example, does this mean with respect to the second law stating that thermodynamically isolated chemical systems approach equilibrium and their entropy increases to a maximum? In the familiar setting of a liter of gas at room temperature, equilibrium of macroscopic features is attained rapidly, and small macroscopic fluctuations such as deviation from chemical equilibrium among a fixed set of molecular species damp out fairly rapidly. By contrast, consider a giant cold molecular cloud in a spiral galaxy with about a hundred million solar masses; ignore gravitational effects and just consider the ongoing complex chemical reactions on the complex dust particles that exist in those clouds. The specific molecular configurations that arise almost certainly include molecular species that are ever unique in the history of the universe. If we may consider unique molecular species as macroscopic features of the cloud, then these fluctuations in macroscopic properties do *not* damp out; rather, they form the nexus for the generation of still new, unique molecular species. A specific cloud, like our biosphere, presumably becomes kinetically trapped into a very special set of complex molecular species that happen to have formed as the cloud evolves.

In short, since the relevant timescale for the ergodic hypothesis to hold is vastly longer than the actual present history of the universe, the macroscopic features of the universe with respect to the specific sets of complex molecules that exist on this planet—in giant cold molecular clouds, and so forth—are kinetically trapped into an infinitesimal subset of those molecular species that might have come into existence in an ensemble of different histories of the universe.

My quip above about the nonergodicity of the universe with respect to species, languages, legal systems, and Chevrolet trucks was not a jest. We noted in the previous chapter that autonomous agents can form hierarchies—hypercycles made of replicators linked in a cycle of mutual benefit are but the simplest case. The symbiotic construction of eukaryotic cells by merging of different bacterial forms to create mitochondria and chloroplasts and perhaps cell nuclei are another case. So too are multicelled organisms such as starfish and ourselves. These hierarchically complex autonomous agents have, do, and will invade an adjacent possible, definable at least at the chemical level, but also on morphological levels, behavioral levels, and beyond. At all levels, the biosphere has been invading a persistent adjacent possible for 4.8 billion years.

Which leads to an odd thought: The indefinite hierarchy upward in complexity is a "sink" where the burgeoning order of the universe constructed by such agents can be "dumped." The biosphere has been doing this dumping for 4.8 billion years.

Now here are some further odd thoughts. There is an absolute zero temperature. You cannot get colder than absolute zero, where the only motions are quantum in nature. Because there is an absolute zero, and the extraction of work via the use of heat differences, as in the Carnot cycle, requires dumping heat from a hotter to a colder reservoir, such "work cycles" require an ever colder sink and are bound to arrest at absolute zero. So, more or less, follows the dreaded "heat death" of the universe.

On the other hand, as we have just seen, the universe is vastly nonergodic at levels of complexity of complex organic molecules upward to autonomous agents co-evolving with one another and beyond. There appears to be no upper bound on this complexity—there is no obvious upper bound that limits this sink, as absolute zero limits work cycles in heat engines. So it may begin to be worth raising the question whether the universe can expand into an adjacent possible for vastly longer periods than the current lifetime of the universe, becoming evermore kinetically trapped, thus evermore specific and more refinely differentiated.

Formalizing the Adjacent Possible

It is helpful to attempt to formalize the concept of the adjacent possible. I will do so using classical physics and the now familiar concept of a $6N$-dimensional phase space. Recall that our particles in the liter box had three positional variables and three momenta, or velocity variables, hence, six numbers per particle. For an N particle system, this is the familiar $6N$-dimensional phase space.

Let's just go ahead and define the classical $6N$-dimensional phase space for a region of real space, including the sphere centered on the Earth out to twice the orbit of the moon and containing the positions and momenta of all particles from the Earth out to twice the orbit of the moon. Physicists always assure us that this makes sense. As usual, we can break this classical $6N$-dimensional phase space into a very large number of tiny "cells," each also $6N$-dimensional. At any moment, the Earth-centered $6N$ system, call it the "Earth system," is in one of these tiny cells, or microstates. Over time, the Earth system—or a larger one including the entire solar system or our galaxy or the local cluster of galaxies—flows from microstate to microstate. By our arguments above and ignoring gravity—a grave mistake in itself—clearly our Earth system has and will flow nonergodically in this phase space for vastly long time periods.

Now, the total adjacent possible to the current microstate of our Earth system is just the total number of microstates that are adjacent to our current microstate. And that number is very large indeed, for it is on the order of 2^{6N}. Say there are 10^{40} particles in the Earth system, a very crude guess, then the total number of adjacent possible microstates is about 2 raised to the 10 to the 41st power. In short, in principle the next state of our chunk of the universe is drawn from among 2 raised to the 10 to the 41st power neighboring microstates.

Now let's define the real adjacent possible. Each point in our current microstate in its classical $6N$-dimensional phase space lies on a specific trajectory that eventually either stays in the current microstate or leaves the current microstate to flow to one particular adjacent microstate. Consider all the points in the current microstate, and for each, draw a red arrow to the neighboring microstate into which that point flows. Then the real adjacent possible is the collection of all neighboring microstates reached by all the red arrows leaving our current microstate.

This is a perfectly fine definition in classical physics. The real adjacent possible from the current microstate might be a single adjacent microstate into which all arrows flow. That would mean that all the points in the current microstate lie on trajectories flowing to the same adjacent microstate.

Or the real adjacent possible might be the case that arrows flow from the current microstate to two adjacent microstates. That would mean that the points in the current microstate can be partitioned into two classes, perhaps lying in distinct regions of the current microstate. One class of points flows to one of the adjacent microstates; the other class flows along trajectories to the other adjacent microstate. But the implication of the existence of two classes of points in our microstate is that its symmetry is broken—the space is broken into two regions, each of which may be compact or be intermixed and intertwined volumes. One volume flows to one adjacent possible microstate; the other flows to the other adjacent microstate.

More generally, we can define the dimensionality of the real adjacent possible with respect to any microstate in the Earth system as the number of adjacent microstates into which flow occurs from somewhere within the current microstate. The dimensionality of the adjacent possible from a given microstate might be as low as 0 or as large as the mathematical number of adjacent microstates, 2 raised to the 10 raised to the 41st power.

Clearly, the larger the dimension of the adjacent possible, the more symmetries have been broken within the current microstate, for it is broken into at least as many volumes internally as the dimensionality of the adjacent possible from that microstate. Of course, in the case of a classical deterministic $6N$-dimensional phase space, the Earth system flows from its current microstate to only one of the real adjacent possible microstates.

It is interesting to remark that at some point if the dimensionality of the adjacent possible of a microstate of the Earth system increases enough, the volume within a microstate corresponding to flow to a specific adjacent possible microstate will become small enough that it must run up against Heisenberg's "uncertainty principle." Then at the quantum level the current microstate can have an amplitude to flow to many of the adjacent possible microstates. Which way the current microstate flows is then no longer deterministic, but a matter of throws of the quantum dice.

Historical Expansion into the Adjacent Possible and Hints of a Law

Let's now ask whether we think that the dimensionality of the adjacent possible of the Earth's biosphere has increased or decreased in the past 4.8 billion years. Consider as a start a liter of living bacteria and a liter of their dead, homogenized molecular components. In the living system, small fluctuations in chemical concentrations within cells are turning myriad genes on and off in the complex system of genetic regulatory networks known to exist in bacteria.

An example of a small section of the genetic network is the lactose operon in *E. coli*. The lactose operon contains three structural genes, that is, genes encoding proteins, and two nearby small sequences of DNA, called a "promoter" and an "operator." The promoter and operator act to regulate the transcription of the structural genes into RNA. Normally, a repressor protein synthesized from a distant gene binds to the operator, blocking transcription of the structural genes from the promoter. In the presence of lactose, however, the lactose binds to the repressor protein and changes its configuration such that the repressor leaves the operator, freeing it. In that condition, other proteins bound at the promoter are able to transcribe the structural genes of the lactose operon. Included among these is the enzyme beta-galactosidase. Beta-galactosidase metabolizes lactose. Thus, the cell normally does not make the beta-galactosidase enzyme, yet in the presence of the metabolite for which that enzyme is required, the lactose operon works to turn on synthesis of the very enzyme that metabolizes lactose. But the fact that small changes in the internal lactose concentration within E. coli turn on synthesis of the lactose operon, including the beta-galactosidase that metabolizes lactose, is precisely the kind of "threshold event"—the operon switches on or does not switch on —that constitutes the breaking of symmetries in the current microstate of the liter of living bacteria. Mathematically, the thresholds become "separatrices" in the chemical state spaces of the bacteria, on one side of which the lactose operon turns on and on the other side of which it does not.

In fact, the lactose operon system is a bistable switch. Once the operon is switched on, one of the three structural proteins is a permease that enhances transport of lactose into the cell. Thus, once the operon is activated, it will remain active, even if the concentration of lactose outside the cell is lowered from an initial high level required to activate the operon to some intermediate concentration. At that intermediate external concentration, the cell can be stably in two states, lactose operon active or lactose operon inactive. Then small fluctuations of internal lactose concentration that cross the internal threshold separatrix concentration can cause the cell to jump from one to the other state of activity and remain in the other state for a relatively long time.

For the liter of living bacteria, the number of adjacent possible microstates is on the order of at least the number of different on-off combinations of activities of genes and metabolic products of which all the genetic regulatory networks of all the cells are capable. It seems obvious that the number of real adjacent possible microstates, hence the dimensionality of the adjacent possible from the current microstate of the liter of living bacteria, is very much larger than the dimensionality of the adjacent possible of the liter of dead and homogenized bacteria at the same temperature.

This simple observation suggests that in the past 4.8 billion years since life arose and autonomous agents began coconstructing a biosphere linking exergonic and endergonic reactions into a diversifying web of ways of making a living, as the molecular diversity, species diversity, and behavioral diversity has increased, the dimensionality of the adjacent possible of the biosphere as a whole, and most typical chunks of it, has increased dramatically.

With respect to my comment above about the broken symmetries eventually hitting the Heisenberg uncertainty limit on the volumes in each microstate, it is probably of more than passing interest that real living entities, cells, do straddle the classical and quantum boundary. One photon hitting a visual pigment molecule can beget a neural response. In short, real living systems straddle the quantum classical boundary. If there is a tendency of coevolving autonomous agents to increase the diversity of alternative events that can occur, then living entities must eventually hit the Heisenberg uncertainty limit and abide at least partially in the quantum realm.

Indeed, the hypothesis that living entities must eventually abut and even transgress the Heisenberg uncertainty limit and abide partially in the quantum realm leads to an intriguing hypothesis. In chapter 10 we will consider "quantum decoherence." This is a quite well-established phenomenon in which the quantum amplitudes propagating along different possible pathways between the same initial and final state can lose phase information. This loss of phase information then prevents the constructive and destructive interference that is the hallmark of quantum phenomena. The loss of the capacity for interference would mark the transition to classical behavior. Many physicists now think that such decoherence constitutes a modern interpretation of the famous "collapse of the wave function" during a measurement event, as posited by the Copenhagen school. The collapse of the wave function converts the propagating superposition of quantum possibilities into an actual, classical event.

The persistent intermingling of quantum and classical phenomena in a living cell might require quantum coherence, but that coherence is widely doubted at the normal temperatures of cells and organisms. On the other hand, persistent intermingling of quantum and classical phenomena might well occur and not require quantum coherence if the timescale of decoherence is close to or overlaps the timescales of cellular-molecular phenomena. Recent calculations suggest that the timescale of decoherence of a protein in water at room temperature might be on the order of 10^{-19} seconds. Thus, it is interesting that proteins and other organic molecules have modes of motion on timescales over many orders of magnitude, spanning from tens of seconds down to 10^{-16} second or less. Thus, the timescale of decoherence is almost the same as the rapid molecular motions in cells. It does not seem totally implausible that cells persistently abide in both the quantum and classical realms, in which the persistently propagating superposition of amplitudes for alternative molecular motions decohere on very rapid timescales and thereby help choose the now classical microstates of proteins and their motions as those proteins couple their coordinated dance with one another to carry out the alternative behaviors that guide a cell in its next set of actions, its adjacent possible. In short, cells may feel their way into the adjacent possible by quantum superpositions of many simultaneous quantum possibilities, which decohere to generate specific classical choices. Such a hypothesis should be testable.

More, at the high risk of saying something that might be related to the subject of consciousness, the persistent decoherence of persistently propagating superpositions of quantum possibility amplitudes such that the decoherent alternative becomes actualized as the now classical choice does have at least the feel of mind acting on matter. Perhaps cells "prehend" their adjacent possible quantum mechanically, decohere, and act classically. Perhaps there is an internal perspective from which cells know their world.

Having now defined the dimensionality of the adjacent possible and noted that the biosphere and universe as a whole is vastly nonergodic, hence, kinetically trapped in a small region of its total space of possibilities, it is fair to wonder whether general laws may govern this nonergodic flow. Given that the dimensionality of the adjacent possible of the biosphere has expanded in the past 4.8 billion years, I want to make the obvious conjecture at a law: Our biosphere and any biosphere expands the dimensionality of its adjacent possible, on average, as rapidly as it can.

I will return just below to think about bounds on this expansion. For, as we will see, it seems reasonable that if the expansion were too rapid, the system would destroy the propagating organization of autonomous agents whose coevolution and increasing diversity is what drives expansion into the adjacent possible and tends secularly to increase that dimensionality. Autonomous agents persistently stumble onto new ways of making a living with one another and exploit those new ways. The biosphere's advance into the adjacent possible is just exaptation over and over again.

In fact, it seems reasonable to think of the "workspace" of the biosphere, that is, what can happen next, as its actual plus its real adjacent possible. It seems likely, and I do conjecture, that the biosphere is expanding its workspace, on average, as fast as it can do so without destroying itself in the process.

And brazen biologist that I am, I begin to wonder whether the universe as a whole in its nonergodic flow might be expanding the dimensionality of its total workspace including its adjacent possible as a secular trend. If so, then since the big bang, the universe persistently diversifies and becomes more complex in such a way that the diversity of different possible next events keeps increasing as rapidly, on average, as is possible. The greater the current diversity of matter, processes, and sources of energy, the more ways there are for these to couple to generate yet further novelty, further symmetry breakings. For this to be correct, time would have to have a directionality toward persistently broken symmetries. And an arrow of time would lie in this directionality. In chapter 10 I return to these issues. There may be grounds to understand why the universe is so complex. In a generalization of our image of a self-constructing biosphere, the universe may construct itself to be as complex and diverse as possible.

If one could ever show such a law, a law in which the diversity and complexity of the universe naturally increases in some optimal manner, that would be impressive. Some fourth law of thermodynamics? An arrow of time? In short, one intriguing hypothesis about the arrow of time is that the nonergodic universe as a whole constructs itself persistently into an expanding adjacent possible, persistently expanding its workspace. This is in sharp contrast to the familiar idea that the persistent increase in entropy of the second law of thermodynamics is the cause of the arrow of time. But the second law only makes sense for systems and timescales for which the ergodic hypothesis holds. The ergodic hypothesis does not seem to hold for the present universe and its rough timescale, at levels of complexity of molecular species and above. Perhaps we are missing something big, right in front of us.

The nonergodicity of the universe as a whole and the biosphere in particular is interesting from another point of view. History enters when the space of the possible that might have been explored is larger, or vastly larger, than what has actually occurred. Precisely because the actual of the biosphere is so tiny compared to what might have occurred in the past 4.8 billion years and because autonomous agents can evolve by heritable variations that induce propagating frozen accidents in descendant lineages, the biosphere is profoundly contingent upon history.

Bounds on the Growth of the Biosphere's Adjacent Possible

The first point to discuss about critical limits to the growth of the dimensionality of the adjacent possible is that major extinction events have occurred. Presumably the molecular diversity and certainly the species and behavioral diversity of the biosphere were devastated during such events. There are two schools of thoughts on these extinctions: the catastrophists and the endogenists. The catastrophists point to meteors, like the monster that hit off the coast of the Yucatan at the end of the dinosaur era, presumably, but not certainly, causing their extinction. The endogenists, including me, admit some big rocks plummeted but note the power law distribution in the size of extinction events, with many small ones and few large ones, and see in these signs self-organized criticality models, discussed in the next chapter, in which many small and few large extinction events arise from the endogenous coevolutionary behavior of ecosystems.

Particularly if we who favor endogenous dynamics are correct, I am precluded from arguing that biospheres endogenously always increase their adjacent possible. For I, among others, predict endogenous biosphere shenanigans among coevolving autonomous agents as the causes of small and giant extinction events. If there is any trend to increase the adjacent possible, it can only be a secular trend. More, any such expansion must ultimately be limited on Earth. One cannot have fewer than one member per species disporting themselves on, in, and around this globe. Each species member does occupy a hunk of three-dimensional space and the planet is only so big.

But I am more concerned with a probable endogenous self-regulation of any advance into the adjacent possible. If that advance into the adjacent possible were to take place too rapidly, it would tend to destroy the organismic propagating organization that is the expansion's foundation and persistent wellspring.

Recall the concept of a supracritical chemical reaction system. Such systems persistently generate molecular novelty. As we saw in discussing the origin-of-life problem, at a critical diversity of molecular species and potential catalysts, a phase transition occurs in which the catalyzed reactions form a giant connected component. Molecular species flow from the founder set actual into a persistently expanding adjacent possible.

I am fond of telling the Noah's Vessel experiment, hypothetical though it is. I ask, thereby, whether the biosphere is supracritical. Take two of every species, all hundred million of them, male and female, normalizing a bit for mass (so you have small bits of hippos and elephants per fly). Dump them all into a large blender and homogenize the hell out of them, breaking all tissue and cell boundaries, spilling out the stuff of life into a common, homogenized liquor.

The small molecule diversity in the blender is presumably on the order of billions, the protein and polymer diversity is on the order of hundreds of trillions, thus 10^{14}. Assuming that any pair of molecular species can undergo at least one two substrate–two product reaction, the total number of reactions is, as noted above, the square of the molecular diversity, so is about 10^{28}. If the probability that any one protein species catalyzes any one reaction is, say, one in a trillion, or 10^{-12}, then the expected number of catalyzed reactions is just the product of the number of reactions times the number of potential protein catalysts, divided by the probability that a given protein catalyzes a given reaction. This yields 10^{28} reactions times 10^{14} proteins divided by 10^{12}, which equals 10^{30}. In short, virtually all possible reactions will be catalyzed by something. Indeed, on average, each possible reaction will find 100 different protein catalysts. A vast sustained explosion into the adjacent possible would occur. Ergo, the biosphere is supracritical. More precisely, the biosphere would be supracritical if all molecular species could be in effective contact with one another on short timescales. But all molecular species do not come in contact with one another willy-nilly, for molecular species are packaged into cells.

It is critical to note that individual cells are *not* supracritical. The crude argument says that a cell's metabolism has about 1000 organic molecules. Consider squirting a novel molecule, Q, into a cell. Presumably Q can be one member of two substrates with each of these 1000 organic molecules. Thus, addition of Q to the cell affords about 1000 novel reactions. Let the protein diversity of the cell be bounded by the human number of genes at 100,000. Any such protein has evolved for some tasks, but may contain molecular nooks and crannies that can serve as novel catalytic sites. The expected number of novel catalyzed reactions due to the presence of Q is given by the product of the number of potential protein catalysts times the number of novel reactions made available by injection of Q into the cell, divided by the probability that any protein catalyzes any given reaction. The product of potential catalysts (the proteins) and the 1000 reactions is 100,000,000, or 10^8. The best current guess at the probability that a randomly chosen protein catalyzes a randomly chosen reaction comes from the probability that a monoclonal antibody canalyzes a reaction and, as discussed above, is about one in a billion. If so, the expected number of the 1000 novel reactions that will be catalyzed when Q is squirted into a cell is 100,000,000 divided by 1,000,000,000, or 0.1. Since 0.1 is less than 1, on average, no chains of novel reactions are catalyzed and cells are subcritical.

If the probability of catalysis is one only in a trillion (as used in the calculation to see if the biosphere as a whole is supracritical), rather than one in a billion, then cells are even more deeply subcritical. With catalysis only one in a trillion, the expected number of catalyzed reactions in a cell upon addition of Q is .0001.

So cells are subcritical. It is a very good thing indeed that cells are subcritical. If cells were supracritical, they would forever generate molecular diversity internal to themselves. Many of the novel molecular species would poison the cell. In short, cells must remain subcritical and cells in communities must remain subcritical, or else the rate of generation of molecular diversity would overwhelm the capacity of natural selection to winnow out the winners from the losers. Everything would die. All propagating organization in the biosphere would rip itself apart in a torrential, if brief, burst of molecular creativity.

In short, the adjacent possible would explode rapidly, but everything around would bite the dust. If cells were supracritical, propagating organization would poison its own propagation.

The fact that cells almost certainly are not supracritical and that the biosphere as a homogenized whole, via the Noah's Vessel experiment, clearly is supracritical, means that the fact that each cell is somewhat isolated from the other and that each has bounded molecular diversity is not an accident. Were it not so, we would not be here.

But what of a microbial community? Consider such a community with N species. As the diversity of species increases, the total molecular diversity of the community increases. At some point, the community as a whole might become chemically supracritical. A novel molecular species, Q, introduced into one species would be sequestered, leave the cell unchanged and be taken up by other cells or lost in the soil, or undergo a reaction to form a known or a novel species. At some diversity of species—N and some rich onslaught of novel molecular species, Q, R, S , . . . —the community will become supracritical. At that stage, molecular diversity in the community increases rapidly. If concentrations of novel molecular species are high enough, say nanomolar or picomolar, then some of the N species will be poisoned. The species diversity of the local community will fall. Presumably, this process can suffice such that the diversity of a causally connected local community falls sufficiently low that the community is not supracritical.

The inverse argument allows the diversity of the community to increase by immigration or mutation of current members. This suggests a possible tendency of local communities to move toward the subcritical-supracritical boundary.

In short, an endogenous process almost certainly limits the rate of generation of molecular diversity such that cells and local communities are not supracritical. The rate of generation of molecular novelty must be sufficiently slow that natural selection can work on heritable variants to pick winners from losers. The point I am making is that it seems reasonable that endogenous processes in local communities gate the rate of exploration of the molecular adjacent possible, keeping it slow enough that natural selection can persistently pick current winners from losers. If so, the biosphere gates its own rate of entry into the molecular adjacent possible. On these arguments, the biosphere may advance into the adjacent possible as fast as it can get away with doing so.

Indeed, it is helpful to frame the current discussion as a generalization of a famous phase transition discussed by Manfred Eigen and Peter Schuster called the "error catastrophe." Eigen and Schuster were considering a population of replicators, say viruses or bacteria, evolving on a fitness landscape with many peaks of high fitness, valleys of low fitness, and ridges. In general, if the mutation rate is low enough, a population located at one point on the landscape will have a few mutants, one or two of which are fitter. These will replicate faster than the less fit cousins, eventually replacing them, so the population as a whole will move to the new point of higher fitness on the landscape. If the process is continued, the population will climb steadfastly uphill to a local fitness peak and remain in its vicinity.

But if the mutation rate is then gradually increased, at some point the population "melts" off the fitness peak and wanders away across the fitness landscape. This melting is the error catastrophe phase transition. Eigen and Schuster elegantly relate the known mutation rates of viruses to the sizes of their genomes and show that viruses are close to but below the error threshold where selection can still overcome the melting. Bacteria, which are metabolically far more complex that viruses, are even more conservative than viruses; their mutation rate is well below the error catastrophe. It is not known why bacteria and higher cells have a mutation rate so far below the error catastrophe. Perhaps were the mutation rate of bacteria higher their communities would become supracritical. And that would be lethal.

Thus, the bounding of mutation rates and community diversity suggests that cells and communities avoid being supracritical, which in turn bounds and gates the entry into the adjacent possible by any local community. On the other hand, the biosphere as a whole is comprised of many different local communities. The rate of exploration of the adjacent possible globally must be bounded such that a generalized Eigen-Schuster error catastrophe, melting the population or community away from adequate organization to survive and propagate, does not occur.

All of this suggests the hypothesis that a biosphere expands into the adjacent possible, as a secular trend, about as fast as it can get away with such exploration, subject to the requirement that selection must on average be strong enough and fast enough to slightly more than offset the rate of exploration of novelty.

It is interesting that the same feature may occur in the economy as a whole. We hear of future shock. Roughly, what we fear is that the rate of technological change will overwhelm us. But will it? Or is there a self-regulating mechanism that gates our rate of entry into the technological adjacent possible?

The latter, I think. Consider this: Why does an innovation get itself introduced? Because someone thinks he or she can make money introducing that innovation. But if the person or firm making the innovation and introducing it to the global or village markets faced a product life cycle that was so very rapid that neither they nor others in the economy could absorb the innovations and make livings, the firms in question would go broke. We will only broach the technological adjacent possible at that rate at which we can make a living doing so. We gate our entry into the technological future.

Thus, it appears that the biosphere and the econosphere have endogenous mechanisms that gate the exploration of the adjacent possible such that, on average, such explorations do successfully find new ways of making a living, new natural and business games, at a rate that can be selected by natural selection, or its economic analogue of success or failure, at a rate that is sustainable. It is a further plausible hypothesis that the rate of exploration of the adjacent possible endogenously converges to the rate that is maximally sustainable.

I close this chapter with a surprising calculation and conjecture by Harold Morowitz, a biophysicist at George Mason University. Morowitz considers the atoms that form organic molecules, C, H, N, O, P, S: carbon, hydrogen, nitrogen, oxygen, phosphorus, and sulphur. He then considers molecules of these elements and the kinds of bonds that can form between the elements. There are two kinds of bonds, chain-terminating bonds, like a terminal hydroxyl group, -OH, and chain-extending bonds, like $C=C$. Chain-extending bonds will tend to make a molecular system with many kinds of molecules even more diverse. Chain-terminating bonds will tend to limit the diversity of molecular species that can form.

Morowitz then considers the equilibrium ratio of chain-extending bonds to all bonds, chain-extending plus chain-terminating, that would occur as a function of temperature, or equivalently, the energy per unit volume of the system. At very high temperatures, the system is a plasma, and chain-terminating bonds are vastly more predominant at equilibrium than chain-extending bonds. At very low temperatures, chain-terminating bonds predominate overwhelmingly.

If one plots the equilibrium ratio of chain-extending bonds to all bonds as a function of temperature, or energy per unit volume, the curve goes up, reaches a single peak, then trends downward. Thus, there is an optimal temperature, or energy per unit volume, where the equilibrium ratio of bonds maximizes the ratio of chain-extending chemical bonds to all bonds. Morowitz's remarkable conclusion is that the maximum of this curve, where chain-extending bonds are as abundant as possible at equilibrium, corresponds quite closely to the average energy per unit volume of the biosphere of the Earth!

Even if the calculation is crude, I find this result deeply interesting. It suggests that, somehow, the biosphere has achieved an energy per unit volume such that at equilibrium chain-extending bonds are maximized. But this means that the energy requirements to form a biosphere of high molecular diversity are minimized!

It is as if the biosphere has managed to get itself to an energy per unit volume that permits the maximum expansion of molecular diversity and, thus, the maximum expansion of the workspace of the biosphere. How might that happen? We can consider the plausible versions of the Gaia hypothesis in which a simple model planet has black or white daisies. The relative abundances of these tunes the albedo, or reflective power, of the planet, hence, the fraction of solar energy absorbed by the biosphere. Simple models show that the ratio of black and white daisies can evolve to maximize their joint fitness, thereby tuning the energy per unit volume of the biosphere.

Morowitz's calculation should be taken cautiously. My conclusions based on his almost back-of-the-envelope calculation should be taken even more cautiously. The arguments are cogent but unestablished. On the other hand, the biosphere has exploded in molecular diversity, the workspace of the biosphere has expanded, the adjacent possible of the biosphere has expanded. It is certainly interesting if the energy per unit volume of the biosphere is roughly that which makes this expansion as energetically inexpensive as possible for the autonomous agents coevolving with one another, exapting to new forms of making a living playing natural games, as we coconstruct our biosphere.

In summary, I hold out the very interesting conjecture that the biosphere as a whole evolves as a secular trend to expand its workspace, including the dimensionality of the adjacent possible as fast as is sustainably possible. If so, we have broached a tentative law for any biosphere. In the next chapter I discuss three further candidate laws. Treat all of them with great caution. It is enough if at this stage we can even begin to formulate tentative laws for all biospheres. A general biology is hardly begun, let alone explored.

Chapter 8

CANDIDATE LAWS FOR THE COCONSTRUCTION OF A BIOSPHERE

F OR THE PURPOSE of our discussion here, grant that molecular autonomous agents propagate organization and evolve by the roughly familiar Darwinian aegis of mutation and selection. These agents—coevolving with one another, discovering displacements from equilibrium that can be used to accomplish work, making records of such sources of energy, then linking those exergonic reactions to endergonic reactions—are the means by which our biosphere has come into being, actually coconstructed by the activities, accidents, striving, and failures of these autonomous agents, exapting persistently into their adjacent possible.

Yes. But how does a biosphere get itself constructed? Are there laws? Are there laws that might hold for any biosphere? Laws of a general biology, wherever autonomous agents swirl into existence and change forever the begetting of the universe?

No one knows. Yet is seems reasonable to expect such laws and honorable to begin, even now, to seek them. At worst we will be wrong. Rather more stunningly, we may be right. It is surely enough if at this early state we can even begin to formulate candidate general laws. Our efforts will only improve over time.

In the present chapter, I consider four candidate general laws for any biosphere. Because the science is more advanced than some of the material in the previous chapters, I will be able to describe it in somewhat more detail. That which is more worked out is, I hope, a signature of how the glimmered science of the past seven chapters may develop.

Coevolutionarily constructible communities of molecular autonomous agents may evolve to four apparently different phase transitions:

LAW 1.   Communities of autonomous agents will evolve to the dynamical "edge of chaos" within and between members of the community, thereby simultaneously achieving an optimal coarse graining of each agent's world that maximizes the capacity of each agent to discriminate and act without trembling hands.

LAW 2.   A coassembling community of agents, on a short timescale with respect to coevolution, will assemble to a self-organized critical state with some maximum number of species per community. In the vicinity of that maximum, a power law distribution of avalanches of local extinction events will occur. As the maximum is approached the net rate of entry of new species slows, then halts.

LAW 3.   On a coevolutionary timescale, coevolving autonomous agents as a community attain a self-organized critical state by tuning landscape structure (ways of making a living) and coupling between landscapes, yielding a global power law distribution of extinction and speciation events and a power law distribution of species lifetimes.

LAW 4.   Autonomous agents will evolve such that causally local communities are on a generalized "subcritical-supracritical boundary" exhibiting a generalized self-organized critical average for the sustained expansion of the adjacent possible of the effective phase space of the community.

Candidate Law 1: The Dynamical Edge of Chaos

Molecular autonomous agents, for example, free-living cells, are parallel-processing molecular dynamical systems. A bacterium such as *E. coli* has on the order of three thousand structural genes. The diversity of molecular species in *E. coli* includes perhaps a thousand small molecules in metabolism, the genes, RNA and protein species, lipids, large carbohydrates, and so forth. For the sake of argument, let's say there are about five thousand molecular species in *E. coli*. Perhaps the number is larger.

A cell is a parallel-processing dynamical system. That is to say, the cell carries out a wide variety of molecular activities, including the turning on and off of transcription of genes into RNA; the processing of that RNA into mature messenger

RNA; the translation of that RNA into proteins, the activities of many of those proteins as enzymes to catalyze reaction, the modification of the activities of enzymes by chemical events such as phosphorylation and dephosphorylation; the building of structural components such as bilipid membrane, and microtubule assembly and disassembly; and the construction of proteins and other receptors. These receptors are located transmembrane at the cell boundary and elsewhere in the cell, including on nuclear membranes of eukaryotes, such that signal molecules can be detected and responded to.

So, lots of activities are going on all the time, in parallel, in your typical *E. coli*, yeast, or your own pancreatic cells. The proper conceptual framework to think about all this activity is the "state space" of the system. If we ignore geometry, that is, the locations of molecules relative to one another in the cell—which is a big idealization—then the state space of a cell consists of a list of all the molecular species and their "activities" or "concentrations."

In the following, I will focus on the behavior of the genetic regulatory network. It has been known since the seminal work of Jacob and Monod in 1961 and 1963, work for which they won the Nobel Prize, that genes can turn one another on and off. In more detail, the protein made by one gene can diffuse in the cell and bind to a DNA site, called a "cis acting site," near a second gene. Genes that encode proteins are the structural genes; the binding of the protein, or several proteins at a set of nearby cis sites, can turn the second structural gene on or off. More generally, the binding of diffusible factors, called "*trans* acting factors," to cis sites, can tune graded rates of transcription of the nearby structural gene.

The human cell is estimated to have about eighty to a hundred thousand structural genes and between ten thousand to perhaps a hundred thousand cis acting sites. In general, any structural gene may be regulated by zero to ten different trans acting factors that may bind at one or more nearby cis acting sites. Therefore, the human genomic system is a highly complex web of regulatory connections and interactions by which the activities of genes turn one another on and off, or more generally tune one another's activity.

It is the joint dynamical behavior of such genetic networks, plus the remaining cellular network of proteins and other molecular interactions, that controls cell behavior, including development from the fertilized egg to the adult.

In the past three decades, considerable theoretical insight has been achieved with respect to the expected behaviors of such large genetic regulatory networks. I discuss the relations to experimental evidence briefly below.

Boolean Networks

To take a very simple case, we consider the N genes of a cell and idealize further to imagine that at any moment in time a gene is either actively transcribing into RNA, with active = 1, or it is not transcribing, so is inactive, hence 0. Thus, the genes are treated as binary, or "Boolean," variables. A Boolean network is a model genetic network with N binary, or Boolean, genes, each receiving regulatory inputs from some among the N genes and each governed by a Boolean function on its inputs telling the activities of its inputs for which it should turn on or off.

The Boolean idealization is severe, but it is a very useful place to start. If the human genome has 80,000 structural genes and each can be on or off, then the number of possible patterns, or "states," of gene activity in the human genome is a staggering $2^{80,000}$ or about $10^{24,000}$. That is, a human cell could, in principle, be in any one of a $10^{24,000}$ states of gene activity. Its state space is $10^{24,000}$. There has only been 10^{17} seconds since the big bang. If it took merely a second to turn a gene on or off, then no human cell could have explored more than an infinitesimally tiny fraction of its state space, $10^{-23,983}$, even if it had been chugging along since the big bang.

That cannot be what cells do. Something must confine their "flow" in their state space. And, indeed, what confines their flow is precisely the genetic regulatory network by which genes turn one another on and off.

Very good work shows that such networks can exist in three broad regimes: an ordered regime, a chaotic regime, and near a phase transition between order and chaos. All the evidence suggests that cells have evolved to lie in the ordered regime, fairly near the edge of chaos. Communities of cells may lie even closer to the edge of chaos. The hypothesis that cells and communities of cells lie in the ordered regime near the phase transition to chaos is candidate law 1.

To understand this candidate law, I need to describe to you the structure and behaviors of model genetic networks. Thereby we can characterize the ordered, chaotic, and edge-of-chaos regimes.

To take a very simple case, we consider a cell with three genes, A, B, and C. In this simplest Boolean idealization, there are three genes, each of which can be on or off, hence, there are two raised to the third, or eight possible states of gene activities: (000), (001), (010), (011), (100), (101), (110), (111), where the ordering of the three symbols stands for the activity states of ABC, respectively.

The most general description of a dynamical system consists in specifying its state space, then identifying for each state which state or states it changes into. For a deterministic dynamical system, each state changes into a unique successor state. For a nondeterministic system, single states can change to two or more successor states. Which of the successor states is chosen in the nondeterministic system is given by some random process such as flipping a coin.

Figure 8.1a shows an arbitrary deterministic state space among the eight states of three genes, A, B, C. For each state, I have chosen its successor state at random.

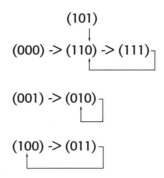

Figure 8.1a shows several characteristic features of these very simple Boolean dynamical systems. First, note that the system is parallel processing. More than a single gene changes its activity value from 0 to 1 or 1 to 0 on many of the state transitions. Next, there is a finite number of states, here, eight. Each state has a unique successor. Over time, if the system is released from any initial state, it will follow a trajectory of states through state space. Since there is a finite number of states, eventually the trajectory must hit a state previously encountered on the trajectory. But the system is deterministic, thus once the trajectory reenters a state previously encountered, it will follow a recurrent loop of states in state space, called a "state cycle."

In general, the length of a state cycle can be a single state that reenters itself, a "steady state," or all of the states in the state space may lie on a single, long cycle that traverses all the states of the state space or state cycles may be any length between these two limits.

A second typical property of such a parallel-processing Boolean system is that more than one state cycle may exist. In the present example, three state cycles exist in the state space.

State cycles are called "attractors" because they typically attract the flow of other states into themselves. This is shown in the first state cycle, where the states (000) and (101) flow into the state cycle, but are not on it. These two states are called "transients." Transient states are encountered on trajectories flowing to state cycle attractors, but are not encountered again once the attractor is reached, assuming no perturbations occur to the system.

The set of states flowing into a state cycle attractor plus that state cycle is called the "basin of attraction" of the attractor.

The set of attractors are jointly the asymptotic long-term alternative behaviors of the network. If released from any initial state, the system ultimately winds up cycling on one of its attractors.

Thus in discrete-valued deterministic networks—here, binary ones—each state lies in a single basin of attraction, so the basins of attraction partition the state space into disjoint sets of states.

The simple example of Figure 8.1b allows us to show another feature of such synchronous Boolean networks—here, synchronous means that all the binary variables change value at the same clocked moment. Since each state has a unique successor state, we can write a table of all the states and for each, its unique successor.

Figure 8.1b shows the state transitions for each state, at time T, to the state it transforms to one clocked moment later, at time $T + 1$.

But Figure 8.1b also shows for each gene, in order (ABC), the Boolean rule, or Boolean function, of the three genes, A, B, C, that turns each on and off as a function of the values of itself and the other two genes.

As it happens, genes A and B have Boolean functions that depend on all three genes, A, B, and C. By examination, however, gene C has a Boolean function that depends only on genes A and C, not on gene B. To say that the activity of gene C depends only on A and C and not B means that once the combinations of activities of A and C are defined at a moment T, the next activity of C at $T + 1$ is independent of whether B is on or off at time T. Indeed, the Boolean function is C = (not A or not C); that is, gene C will turn on at the next moment if at the current moment either A is not active or C is not active or both are not active.

From Figure 8.1c, after simplifying Boolean expressions as we just did for gene C, we can write down the "wiring diagram" of inputs among the three genes. Since

T		T + 1
(ABC)		(ABC)
(000)	→	(110)
(001)	→	(010)
(010)	→	(010)
(011)	→	(100)
(100)	→	(011)
(101)	→	(110)
(110)	→	(111)
(111)	→	(110)

FIGURE 8.1b

| T | T + 1 | T + 1 | T+1 |
ABC	A	B	C
(000)	1	1	0
(001)	0	1	0
(010)	0	1	0
(011)	1	0	0
(100)	0	1	1
(101)	1	1	0
(110)	1	1	1
(111)	1	1	0

FIGURE 8.1c

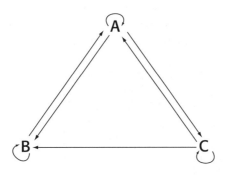

FIGURE 8.1d

A and B depend upon all three genes, each receives a regulatory input from all three genes. Gene C, however, receives inputs only from itself and gene A.

The combination of Figures 8.1d and 8.1c, respectively, shows for each gene the inputs to that gene and the logical or Boolean function by which it turns on and off. Thus, the combination of 8.1c and d is the genetic network among the genes.

Order, Chaos, and the Edge of Chaos

As noted above, thirty years of work by many scientists—initially on synchronous Boolean networks but now generalized to a wider family of model genetic systems, including some where genes can exhibit continuously graded levels of activity as their inputs turn gradually up or down—all show the same simple, general results: There are three broad regimes of behavior. A number of very simple properties of networks, involving connectivity of the network and simple biases on the Boolean functions, control which regime a network lies in.

The Ordered Regime

Figure 8.2 shows a hypothetical movie of a network in the ordered regime. Let the network be released from an arbitrary initial state and flow along a trajectory toward a state cycle attractor. If a gene is rapidly turning on and off, or twinkling, call it green. If the gene is frozen in the active or frozen in the inactive value for a long time, say, fifty state transitions or more, call it red. In particular, once the system is on its state cycle, the twinkling genes that turn on and off on the state cycle would be green and those genes frozen on or frozen off, red.

Initially, just after release from some random initial state, most genes are twinkling, hence green. As the network approaches its state cycle, more and more genes turn red, hence are frozen on or frozen off. By the time the system has reached its state cycle attractor, the majority of the genes are colored red (Figure 8.2a).

Most critically, if one considers the subset of red genes, they form a giant "percolating cluster" whose size scales linearly with the size of the entire network. In effect, the frozen red component is a "frozen red sea," which spans the entire network and typically leaves behind isolated twinkling green islands.

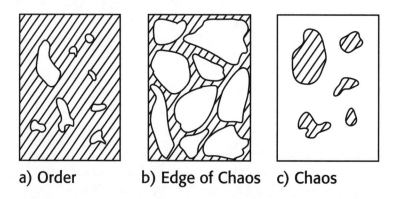

a) Order   b) Edge of Chaos   c) Chaos

FIGURE 8.2 a,b,c Schematic view of the ordered, edge of chaos, and chaotic regime in Boolean networks. In this schematic view, on-off genes are arranged on a two-dimensional square lattice, such that each has inputs from its four neighbors. Start networks at some initial state, let the network run along its trajectory for some time, then call those genes green that are twinkling on and off, and call those genes red that are in fixed states of activity, either frozen on or frozen off. In the ordered regime a connected frozen sea of red genes percolates across the network, leaving behind one or many green twinkling islands. In the present figure, the red frozen sea is denoted by hatch marks, while the twinkling green islands are colored white. In the chaotic regime, a connected twinkling green sea spans, or percolates across the system, leaving none or a few frozen red islands. At the phase transition, or edge of chaos, the twinkling sea is just breaking up into small and large twinkling islands. The intuition is that the most complex, coordinated behavior can occur in networks in the ordered regime near the edge of chaos.

The Chaotic Regime

In the chaotic regime, the same movie shows that the majority of the genes remain green, twinkling on and off (Figure 8.2c). So a vast, twinkling green sea spans the network, typically leaving behind isolated frozen red islands.

The Edge of Chaos

As parameters of the network discussed below are tuned from the chaotic regime toward the ordered regime, the green percolating sea becomes smaller and eventually fragments into two or many isolated green islands. The point of fragmentation of the green sea into green islands constitutes a phase transition from the chaotic to the ordered regime. This phase transition is sometimes called the edge of chaos (Figure 8.2b).

Several critical features distinguish the ordered from the chaotic regime. In the ordered regime, the lengths of state cycle attractors scales polynomially with the number of genes. Remarkably, in the ordered regime near the phase transition to chaos there is evidence for universal scaling in which the number of states on a state cycle scales as the square root of the number of genes. This scaling, which I first discovered over thirty years ago, still staggers me. If the human genome has 80,000 genes, it has a state space of $2^{80,000}$ or $10^{24,000}$ states. Yet if the human genomic system lies in the ordered regime near the phase transition to chaos, it will settle down and cycle among the square root of 80,000, or about 270 states!

Now, 270 states is very very small compared to $10^{24,000}$. The overwhelming order of the ordered regime, I believe, keeps cells from wandering all over their state spaces for eternities beyond eternities. In fact, it takes about one to ten minutes to turn a eukaryotic gene on or off, so it would take a cell from 270 to 2,700 minutes, or from about 4.75 hours to about 48 hours, to traverse its state cycle attractor. This is right in the biological ballpark. For example, the cell division cycle of different human cell types is in the range of 8 to 48 hours.

By contrast, in the chaotic regime state cycle lengths scale exponentially with the size of the network. The deepest one can go into the chaotic regime is to assign at random the successor state for each state. In that case, in general, each gene is a Boolean function of all N genes. In this case, the typical state cycle length is the square root of the number of states in the state space. For the human genome with $10^{24,000}$ states, a typical state cycle length would be the square root, hence $10^{12,000}$. Remember, it is only 10^{17} seconds since the big bang. State cycle attractors of lengths $10^{12,000}$? Not in my body, thank you very much.

In the first two articles I wrote on the subject of random Boolean nets, as long ago as 1967 and 1969, I plotted cell cycle lengths from organisms as diverse as bacteria, yeast, worms, plants, and simple and complex animals. These progressively more complex organisms have progressively more DNA per cell and more genes per cell. If the Boolean net theory is on the right track, and if cell cycles are a reasonable proxy for expected state cycle times as a function of the number of genes, then cell cycle time should scale as a square root function of the number of genes. This prediction is actually pretty much correct. Indeed, a plot of median cell cycle time versus total DNA per cell is a square root function from bacteria to human cells.

But there are caveats: The number of genes in cells may not be proportional to the amount of DNA per cell. Some DNA is "junk." A plausible estimate of the number of genes per cell is now available for many organisms. On this basis, cell cycle time scales somewhere between a square root and a linear—that is, directly proportional—function of estimated genes per cell. So without yet invoking natural selection to tune the structure and logic, the theory of random Boolean nets is already quite close to the data.

A second critical feature that distinguishes the ordered from the chaotic regime is what happens when the activity of a single gene is transiently reversed. I would note that such transient reversals happen all the time in normal development. For example, a single hormone enters a cell, then the nucleus, then binds to a nuclear genetic site and transiently changes the activity of some gene. Typically, the results unleash a cascade of alterations of gene activities. These cascades of alterations guide development and cell differentiation. In the ordered regime, these cascades tend to be smallish. In the chaotic regime, the cascades are typically huge. In real cells, the cascades tend to be smallish. More, we can actually predict their size distribution.

Let's define a gene as "damaged" if after an initial gene has had its activity reversed for a single moment the gene in question ever behaves differently than it would have had the perturbed gene been left undisturbed. In effect, damage shows that perturbation of a gene affects the behavior of the damaged gene. Imagine damaged genes purple. If a gene has misbehaved once, it is purple, whether it stops misbehaving or keeps misbehaving.

Given this definition of damage, we can consider in detail two identical copies of a network, in the same state, running at the same speed. Now pick a model gene at random. If it is on, flip it off. It is off, flip it on. Color it purple since you have damaged it.

Now watch the unperturbed and perturbed copies of the network, and consider purple any gene that ever does something different in the perturbed network compared to the unperturbed network. In general, you will see a purple avalanche spread out from the initially perturbed gene. The purple avalanche will spread out in some way, then must eventually stop. For example, at a maximum, all the genes turn purple. Thus, we can define the size of a given damage avalanche as the number of genes that turned purple, hence misbehaved at least once.

In the ordered regime, something magic happens. If one of the frozen red genes is perturbed, typically no avalanche spreads from that purple gene. If an avalanche spreads at all, it is tiny.

By contrast, if a twinkling green gene in one of the isolated green islands is perturbed and turned purple, a purple avalanche spreads to some or all of the twinkling green genes in that island. But the avalanche stops at the boundaries of the green island since damage avalanches cannot propagate through the frozen red percolating sea.

Because purple avalanches cannot propagate through the frozen red percolating sea, the green islands are functionally isolated from one another. The consequence is that there is a characteristic size distribution of avalanches in the ordered regime and a very different distribution in the chaotic regime.

Figure 8.3a schematizes the distribution of avalanches for networks in the ordered regime very near the phase transition to chaos. The figure plots the logarithm of the size of the avalanche on the x-axis and the logarithm of the number of instances of avalanches of each size on the y-axis. As you can see, the size distribution shows up as a straight line in this log log plot, sloping down to the right. Hence, the distribution is a power law distribution, with many small and few large avalanches of change propagating through the network. In addition, there is a finite cutoff and thus a largest-size avalanche, which seems to scale as a square root function of the total number of genes in the network. Deeper in the ordered regime, the size distribution of avalanches remains a power law, but the slope down

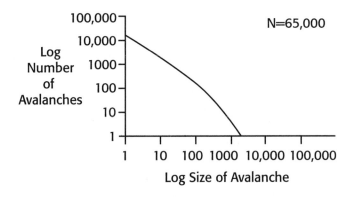

FIGURE 8.3a   Ordered regime has a power law distribution of avalanches of changes in gene activities produced by reversing the activity of a single randomly chosen gene. Model genetic network simulated had N = 65,000 on-off genes, almost equal to the rough 80,000 to 100,000 genes in a human cell. Distribution shows a finite cut off with maximal avalanches about two or three times the square root of the number of genes in the system.

to the right becomes steeper, so there are fewer big avalanches compared to small avalanches.

If a human genome is in the ordered regime near the phase transition to chaos and harbors some 100,000 genes, then the largest avalanches should be about two times the square root of the number of genes, hence 2 x 317, or about 634 genes. This is probably about right. In the fruit fly, *Drosophila melanogaster,* with about 16,000 genes, the largest avalanches should be about twice the square root of 16,000, or 2 x 127, or 254. The largest avalanche I am aware of occurs when the moulting hormone ecdysone acts on the salivary glands and induces changes in about 155 of the "puffs" in the polytene chromosomes. If each puff is a single gene, as most geneticists think, then ecdysone unleashes an avalanche altering the activities of 155 genes.

No comparative data yet show whether the size distribution of avalanches in real organisms is a power law, nor whether the largest avalanches scale as a square root function of the number of genes in the organism. But these hypotheses are fully testable using today's experimental techniques.

Nevertheless, these predictions do roughly fit one's expectation as a biologist. Most genes if perturbed should unleash no avalanches or just small avalanches, fewer genes should unleash larger avalanches, and some modest fraction of the genes at most should be open to alteration by transient alteration in the activity of any single gene. Thus, this typical, or "generic," behavior of parallel-processing networks in the ordered regime closely fits the known data and our informed intuitions.

The chaotic regime contrasts starkly with the ordered regime. Its expected behaviors are not biologically plausible. The chaotic regime differs from the ordered regime for a simple reason. In the chaotic regime, the twinkling green sea percolates. If a single green gene is perturbed and turned purple, that perturbation usually unleashes a purple avalanche that spreads through much of the percolating green sea. Huge damage avalanches are unleashed by single gene perturbations.

These huge avalanches are exactly the signature of the famous "butterfly effect" seen in the weather, where a small initial change can have large-scale consequences. In short, the spreading purple avalanches constitute "sensitivity to initial conditions." On the other hand, it is important to distinguish between low-dimensional chaos, characterized by three or four variables governed by three or four equations, and the high-dimensional chaos, shown in large-model genetic networks with tens of thousands of gene variables. In high-dimensional networks of genes modeled as binary variables, chaos shows up as the enormous avalanches of damage that spread from one to many of the variables of the model network.

Figure 8.3b schematizes the size distribution of avalanches in the chaotic regime. Unlike the ordered regime, there is a spike of huge avalanches where 30 to 50 percent of the genes are damaged. In a cell, this would correspond to a hormone

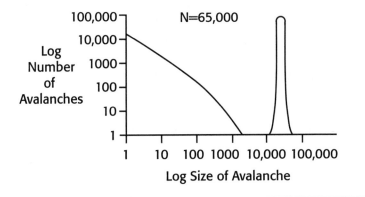

FIGURE 8.3b Chaotic regime has, in addition to a power law distribution of avalanches, a large spike of vast avalanches affecting up to 30–50 percent of the genes. In the chaotic regime, the size of the largest avalanches of change scales with the number of genes in the network, here 65,000, rather than with the square root of the number of genes in the network.

changing the activity of a single gene and 30,000 to 50,000 genes downstream changing their activities. This does not happen.

As in the ordered regime, however, in the chaotic regime there is also a power law distribution of small avalanches, present in addition to the vast avalanches that rocket through the green sea. Presumably, these small avalanches occur when green genes near the filigreed "coasts" of red frozen islands are perturbed and the purple avalanche is trapped on the fingers of the red beaches.

A third feature, a convergence versus divergence along flows in state space that characterizes the ordered versus chaotic regime, is perhaps the most important to our future discussions. In the ordered regime, initially nearby states lie on trajectories that tend to converge in state space. In the chaotic regime, initially nearby states tend to lie on trajectories that diverge in state space. At the edge of chaos, initially nearby states tend to lie on trajectories that neither converge nor diverge in state space.

These behaviors are conveniently shown in a recurrence map, which I call a "Derrida curve" since physicist Bernard Derrida of Saclay, France, first showed it to me. In 1985 Derrida and Pomeau were also the first to find analytic proof of the ordered and chaotic regimes and the phase transition between them.

Consider two states of a Boolean network with five genes and the successors to each of those states:

$$\text{state 1 (00000)} \rightarrow \text{(00110) state 1}'$$
$$\text{state 2 (00001)} \rightarrow \text{(10101) state 2}'$$

We can define the "Hamming distance" between state 1 and state 2, the number of binary variables by which the two states differ. Here the Hamming distance is 1, since gene 5 is 0 in state 1 and 1 in state 2. In addition, we can define the normalized Hamming distance, Dt, between these two initial states at time t as the fraction of binary variables by which they differ. Hence, Dt is $1/5 = .2$. Similarly, we can define the normalized Hamming distance between the two successor states at time $t + 1$. In the case above, $D(t + 1) = 3/5 = .6$.

Then we can compare Dt and $D(t + 1)$ and ask if the initial distance at time T decreased or increased at time $T + 1$. In the current case, the initial distance increased from .2 to .6 at the next moment, $T + 1$. In short, the two initially nearby states, differing in the activity of a single gene, spread further apart one moment later. Indeed, this spreading is the first time step of the spreading of a purple avalanche of damage.

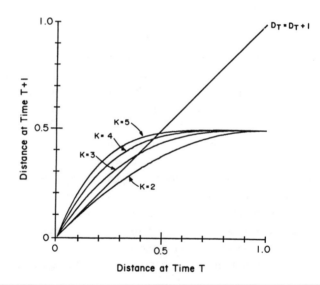

FIGURE 8.4 The Derrida curve, a recurrence relation showing the expected distance, Dt+1 between two states at time T + 1 as a function of the normalized distance, Dt between two states a moment earlier. The main diagonal, Dt = Dt+1 shows the conditions under which two initial states lie on trajectories that neither diverge nor converge in state space. For values of K, the number of inputs per gene, greater than 2, the Derrida curve lies above the main diagonal for small initial distances between initial states, Dt. This corresponds to the first step in an expanding avalanche of damage, and is a signature of chaotic behavior and sensitivity to initial conditions—the butterfly effect. For K = 2 or less, the Derrida curve is below the main diagonal for all initial distances, Dt, corresponding to convergence in state space. K = 2 is the phase transition to chaos.

To characterize Boolean networks with respect to the ordered and chaotic regime, it is convenient to take thousands of random pairs of states at different initial distances, Dt, where Dt can vary from 0.0 to 1.0. For each pair of initial states, run the Boolean network forward one moment, discover the two successor states of the two initial states, and compute $D(t + 1)$ for that pair of initial states. Average the $D(t + 1)$ values for the thousands of pairs of initial states at each initial distance, Dt. The typical results are shown in Figure 8.4.

Figure 8.4 is a recurrence map, with Dt shown on the x-axis, $D(t + 1)$ shown on the y-axis. Thus, for an initial pair of states at $Dt = .1$, if the successor states have spread apart, say to $D(t + 1) = .2$, a dot in the xy plane at $x = .1$, $y = .2$ records this event. The averaged set of these dots for the thousands of pairs of initial states at all different initial distances is the average recurrence map for the network.

In Figure 8.4, the main diagonal, running at a 45-degree angle from the lower-left corner, which corresponds to $Dt = 0$ and $D(t + 1) = 0$, shows the condition where $D(t + 1) = Dt$. If a dot lies on the main diagonal, then the distance between the initial states is the same as the distance between the successor states. The two initial states lie on trajectories that neither diverge nor converge in state space.

In the chaotic regime, nearby states lie on trajectories that diverge further apart in state space, so the recurrence map lies above the main diagonal for small initial distances. For large initial distances, even networks in the chaotic regime have the property that states tend to lie on trajectories that converge. The degree to which the Derrida recurrence curve lies above the main diagonal for small Dt is a measure of how deeply into the chaotic regime the network lies. Deep into the chaotic regime, nearby states, hence small Dt, diverge swiftly. Thus, the recurrence curve is well above the main diagonal for small values of Dt.

By contrast, in the ordered regime, as shown in Figure 8.4, the Derrida recurrence curve is below the main diagonal for small Dt, that is, initial states that are close lie on trajectories that converge. At the phase transition between order and chaos, the Derrida recurrence curve begins, at $Dt = 0$, tangent to the main diagonal, then falls below it as Dt increases.

In summary, in the ordered regime, nearby states tend to lie on trajectories that converge in state space. At the phase transition, nearby states tend to lie on trajectories that neither converge nor diverge. In the chaotic regime, nearby states tend to diverge.

As I will shortly discuss below, it is plausible to think that autonomous agents, and communities of autonomous agents, evolve such that they lie in the ordered regime near the phase transition to chaos. A major reason for this intuition is that under such circumstances flow in state space is mildly convergent. In turn, this will allow the autonomous agents to make the maximum number of reliable discriminations and reliable actions, hence, to play the most sophisticated natural games by which to earn their livings.

The three features that characterize the phase transition between order and chaos seem by good numerical evidence to coincide. That is, when parameters discussed below are tuned from the chaotic to the ordered regime such that the green sea is just breaking up into green islands, simultaneously, the Derrida curve changes from the chaotic regime to become tangent with the main diagonal for small values of Dt. And at just this point, state cycle lengths switch from scaling exponentially to scaling polynomially with the number of genes.

At least three simple parameters tune whether networks are in the ordered or chaotic regimes. Therefore it is important that evolution can readily tune whether genomic systems lie in the ordered or chaotic regime by tuning any of these three parameters. Even more important, evolution seems to have done just that and tuned cells into the ordered regime. Since cells are our only example of evolved autonomous agents, the data support my candidate first law that autonomous agents and communities of autonomous agents will evolve to the ordered regime near the phase transition to chaos.

The simplest parameter to tune is the number of inputs, K, per gene. I showed numerically in 1967, and Derrida and Pomeau showed analytically in 1985, that if $K = 2$ or less, networks lie in the ordered regime. Derrida and Pomeau showed that $K = 2$ is the edge of chaos phase transition. For K greater than 2, networks lie in the chaotic regime.

Already this is worth the excited attention I gave it so long ago. For the results show that a network with randomly chosen logic nevertheless behaves with exquisite order. Say, a network of 10,000 genes is constructed at random, with the simple limitation that each model gene have $K = 2$ inputs but that the wiring diagram be chosen at random, and the Boolean function assigned to each gene among the 16 possible Boolean functions of $K = 2$ inputs is also chosen, once and for all, at random, this spaghetti mess of a network with its tangle of 20,000 wires connecting the genes in some mad scramble will straighten itself out.

The system settles down to cycle among about 100 states out of $2^{10,000}$ or $10^{3,000}$! Order for free, I keep saying. Selection need not struggle against all odds to achieve cells that behave with overwhelming order. That order lies to hand for selection's further craftings.

There are two further known parameters that can tune networks from the chaotic to the ordered regime if K is greater than 2. Both are biases on the Boolean functions. Remarkably, real cells show dramatic evidence of one of these two biases, which I call "canalyzing Boolean functions." I discuss this canalyzing bias second.

The first bias is characterized by a parameter Derrida and colleagues called "P." Consider in Figure 8.1c, the Boolean function for gene A. It has five 1 values and three 0 values. The parameter P is defined as the number of instances of the majority value over the full set of cases. Hence, P for gene A = 5/8. For gene B, the

Boolean function has seven 1 values and one 0 value. Its P is 7/8. For gene C there are six 0 values and two 1 values. Its P is 6/8.

By definition, P for a Boolean function can vary from 0.5 to 1.0, when the majority fraction varies from half to all the possible cases. Derrida and colleagues showed that, in general, when $K > 2$, P can be tuned upward from 0.5 to some critical value, Pc, where networks pass from the chaotic to the ordered regime. The critical value of P as a function of K is shown in Figure 8.5a. Universal scaling for cycle lengths as a square root function of the number of genes has been established along the phase transition in the PK plane.

The second bias in Boolean functions are the canalyzing Boolean functions. Consider gene C in Figure 8.1c. If gene A is 0, gene C will be 0 at the next moment no matter what the activities of gene B or C may be. If gene A is 1, gene C may be 0 or 1 at the next moment, depending on the prior state of gene C itself.

Gene C is governed by a canalyzing Boolean function. Canalyzing Boolean functions have at least one input with at least one value that suffices to guarantee the next state of the regulated gene, regardless of the values of all other inputs. By inspection, if A is 0 now, then C is guaranteed to be 0 a moment later. So the Boolean function is canalyzing, and I call A a canalyzing input to C. Note that gene C is also a canalyzing input to gene C.

Look next at the Boolean function for gene B in Figure 8.1c. If gene A is 1, then gene B is sure to be 1 the next moment, regardless of the activities of B and C. So A

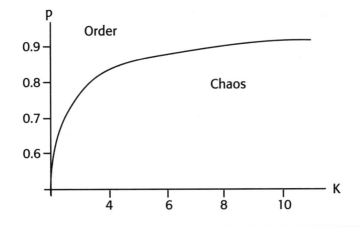

FIGURE 8.5a Networks with more than K = 2 inputs per gene can be shifted into the ordered regime from the chaotic regime by biases on the choice of Boolean functions. This figure shows the phase transition boundary between order and chaos in the *PK* plane, see text.

is a canalyzing input to B. But if B is 0, that too assures that B will be 1 at the next moment. So B is a canalyzing input to itself. And similarly, if gene C is 0, gene B is sure to be 1 the next moment. Gene B has three canalyzing inputs. By contrast, gene A in Figure 8.1c has no canalyzing input, so is not a canalyzing Boolean function. No value of A alone, B alone, or C alone suffices to guarantee the next activity of gene A.

In general, Boolean functions of K inputs may have $0,1,2,\ldots K$ canalyzing inputs. Numerical evidence shows that, for $K > 2$ inputs per gene, a sufficient bias toward a high fraction of genes with a sufficient number of canalyzing inputs drives networks from the chaotic into the ordered regime. Figure 8.5b shows the phase transition curve in the CK plane.

Before turning back to biology, it is essential to stress that the results noted above for synchronous Boolean networks extend to asynchronous Boolean networks and, more critically, extend to a family of model gene networks in which the genes have graded levels of activity. This is important because the on-off Boolean idealization is quite severe. Real genes show graded levels of activities as a function of the concentrations of their trans acting inputs and the bound states of their cis regulatory loci. If our results were fragile in that they depended upon the Boolean, on-off idealization, we could not trust them to inform us about real cells. Glass and Hill have ex-

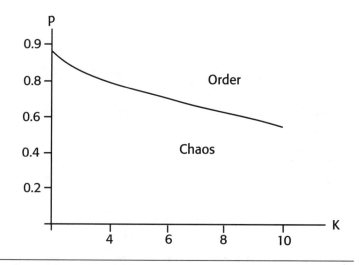

FIGURE 8.5b Networks with more than K = 2 inputs per gene can be shifted into the ordered regime from the chaotic regime by biases on the choice of Boolean functions in favor of high numbers of canalyzing functions. Figure shows the phase transition between order and chaos in the CK plane. Eukaryotic cells seem to use this bias toward canalyzing functions to assure that cells are modestly within the ordered regime (see text).

amined a model with continuously graded levels of gene activity, the "piecewise linear model," and found the same qualitative behaviors. In particular, the same phase transition occurs as a function of K, P, and C. The striking difference, however, is that deep in the ordered regime of the piecewise linear case, the genes of the twinkling green islands settle down to steady states that differ on different attractors. Near the phase transition to chaos, the green islands begin to exhibit sustained "limit cycle" oscillations that become chaotic in the chaotic regime.

Thus, there is now good general evidence that the ordered and chaotic regimes and the phase transition between them are deeply characteristic of some enormous class of parallel-processing nonlinear dynamical systems.

The Biology

But what of real cells? We have no conclusive evidence, yet an abundance of telling hints. If I am not yet entirely convinced and if I am—as I am—biased, I nevertheless become increasingly confident that cells, and probably communities of cells, do live in the ordered regime near the edge of chaos. Not only does the evidence point this way, but cells *should* live near the edge of chaos. Why? As remarked already, the intuition is simple. Being autonomous agents, cells must, as individuals living in communities, make the maximum number of reliable discriminations possible and act on them reliably, without "trembling hands." Just inside the edge of chaos seems the ideal place.

Intuitively, slightly convergent flow in state space allows classification, for when two states converge on a single successor state, those two states have been classified as "equivalent" by the network. Slightly convergent flow would seem to allow the maximum number of reliable classifications in the face of a noisy environment. The convergent flow buffers the system against the noise of the environment.

And what of trembling hands? No point making superb discriminations, seeing the stag deer, drawing your bow, aiming the arrow, then shooting yourself in the foot. Again, slightly convergent flow in state space to buffer external and internal noise seems ideal.

So what about cells? My colleagues Steven Harris, Bruce Sawhill, and Andrew Wuensche, and I have carried out work over the past several years that has analyzed actual gene regulatory rules for eukaryotic genes drawn from a variety of eukaryotic organisms—yeast, *Drosophila*, maize, mouse, and so forth. The results show a very strong statistical bias in favor of genes governed disproportionately by canalyzing Boolean functions. When we have constructed model networks with the observed bias toward canalyzing functions, such networks lie modestly in the ordered regime by the Derrida curve and other criteria noted above.

It all began at the Santa Fe Institute several years ago. Steve Harris, a molecular biologist from Texas was visiting. I told him about canalyzing functions. "Never," said Steve. Seeing my opportunity I replied, "You have a good genetics library, want to read a bunch of papers and analyze the transcription rules of genes with three or four or five known regulatory inputs?"

I didn't think Steve would say yes, for the reading of over a hundred papers in the subsequent years, and cataloging the detailed results, was going to be a substantial task.

"Sure," he replied.

Some months later, Harris called. "Hey, the results for genes with $K = 3$ inputs look interesting! There is a bias toward canalyzing functions."

"Never," I said.

"I'll send the data," was the reply.

Steve had carefully read about sixty papers on regulated genes with $K = 3$ known inputs, where the data was available at the level of actual binding of trans acting factors to cis sites and the turning on of transcription. A gene with $K = 3$ known inputs has, in the Boolean idealization, 2 to the 3rd, or 8, possible on-off states of those inputs, as we have seen. In virtually all the cases used, Steve had good data for all eight input states. He warranted the Boolean idealization had its problems, but found that in many cases the response of a gene was nonlinear to its inputs. Thus, gene A might be turned on 10 percent by factor 1 and 5 percent by factor 2, but 95 percent by both factors at the same concentration. It looks like the Boolean "and" function, where the regulated gene is "on" at the next moment only if both inputs are "on" now.

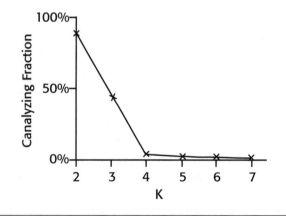

FIGURE 8.6 The fraction of Boolean functions that are canalyzing on one or more inputs decreases from a maximum of 87.5 percent to 4 percent then less, as K increases from 2 upward. The number of Boolean functions of K inputs is 2 raised to the 2 raised to the K power, hence 16 for K = 2, 256 for K = 3, over 65,000 for K = 4, and so on.

We need some mathematical facts. The number of Boolean functions of K inputs is 2 to the 2 to the K, $2^{(2^K)}$. For $K = 2$, there are 16 Boolean functions. For $K = 3$, there are 256 Boolean functions. For $K = 4$, there are 65,353 Boolean functions. For $K = 5$, there are over a billion Boolean functions.

A Boolean function with K inputs, as noted, can have $0, 1, 2, \ldots K$ canalyzing inputs. But, as K increases, the number of Boolean functions that are canalyzing at all, on 1 or more inputs, declines dramatically, as shown in Figure 8.6. In particular, 87.5 percent of the 16 Boolean functions of $K = 2$ inputs are canalyzing. But only 43 percent of the 256 Boolean functions of $K = 3$ inputs are canalyzing. Only 4 percent of the 65,353 Boolean functions of $K = 4$ inputs are canalyzing. Only less than 1 percent of the billion or so Boolean functions of $K = 5$ inputs are canalyzing.

This shift means that we can test if there is a bias in sampled eukaryotic genes. Indeed, in more detail, among the 256 $K = 3$ Boolean functions, 43 percent have no canalyzing inputs and a decreasing fraction of the Boolean functions have 1, 2, or 3 canalyzing inputs, as you can see in Figure 8.7.

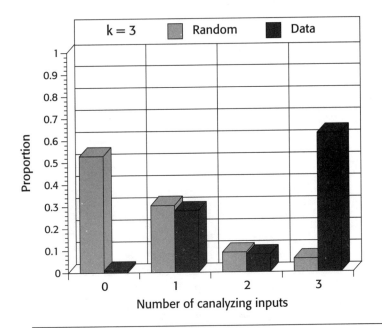

FIGURE 8.7 Comparison of the distribution of numbers of canalyzing inputs for 53 eukaryotic genes with K = 3 known regulatory inputs, compared to the expected number of canalyzing inputs were genetic control rules sampled at random from among the 256 Boolean functions of K = 3 regulatory inputs. Real regulated genes show a dramatic excess of cases with three canalyzing inputs per gene. The results are strongly statistically significant.

Also plotted on Figure 8.7 is the observed fraction of eukaryotic genes with $K = 3$ inputs. The observed curve is the opposite from the curve expected if $K = 3$ genes were regulated by Boolean rules drawn at random from among the 256 functions. Indeed, fully 65 percent of the observed cases have 3 canalyzing inputs, while the expected fraction would be only 8 peercent if rules were drawn at random.

One does not need fancy statistics, but they readily confirm that the observed distribution is sharply shifted to large numbers of canalyzing inputs per gene. Figure 8.8 shows similar results for genes with $K = 4$ known inputs. Again, the shift toward genes regulated with a high number of canalyzing inputs is apparent and strongly statistically significant. Data for $K = 5$ and $K = 6$ genes shows the same bias, but the cases are too few to be statistically significant.

But there remains analysis to be done. Recall the P parameter. Networks with high P values, where genes are mostly turned on or turned off by their inputs, also lie in the ordered regime. Moreover, there is an overlap but nonidentity between

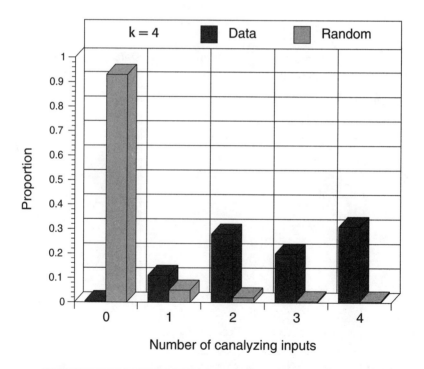

FIGURE 8.8 Similar to Figure 8.7, but with data on 27 eukaryotic genes with $K = 4$ known inputs. Again, regulated genes exhibit a shift toward high numbers of canalyzing inputs compared to a random sample of Boolean functions with $K = 4$ inputs. The results are strongly statistically significant.

the classes of Boolean functions of high P values and Boolean functions with 1 or more canalyzing inputs. When we analyzed our samples of genes, they also had high P values compared to a random distribution of Boolean functions with $K = 3$ or $K = 4$ inputs.

In order to discriminate whether the observed bias was toward high canalyzing inputs or high P values or both, we carried out a "residual analysis." That is, we classified all $K = 3$ Boolean functions into different P classes, $P = 4/8$, $P = 5/8$, $P = 6/8$, $P = 7/8$. Within each P class, some Boolean functions have 0, 1, 2, or 3 canalyzing inputs. Therefore, among all the Boolean functions for $K = 3$ within a given P class, there is some distribution of Boolean functions with 0, 1, 2, or 3 canalyzing inputs. Thus, we asked, within a given P class, if the real genes showed a residual bias toward a high number of canalyzing inputs per gene compared to what would happen if real genes were governed by Boolean rules drawn at random with respect to canalization. The answer for $K = 3$ and $K = 4$ genes is overwhelmingly yes.

In short, if we control for P classes, there is a very strong and very statistically significant residual bias toward high numbers of canalyzing inputs per gene. Conversely, when we controlled for canalyzing input classes and tested for a residual bias toward high P values, there was no sign whatsoever of such a bias. Thus, it appears that evolution has, in fact, tuned the choices of Boolean rules used to govern genes with $K = 3$ and $K = 4$ known inputs, as well as genes with $K = 5$ and $K = 6$ inputs that we have sampled, sharply in favor of a high bias toward usage of Boolean rules that are canalyzing functions.

The main caveat to hold in mind, in addition to misreading the articles or the articles being a nonrandom sample of published data, is that genes governed by canalyzing functions may have more easily detected genetic effects, hence, be noticed and studied. Only future work with randomly chosen structural genes will overcome this source of bias. Despite the caveat, I am quite convinced by the data. In particular, genes governed by high P values would also have easily detected genetic effects, yet there is no such bias in the data.

Tentatively, eukaryotic genes are governed by rules biased toward many canalyzing inputs per gene. Why? Either chemical simplicity or natural selection or both, I think.

Now let's examine the consequences. We know that networks with $K = 3$, $K = 4$, or more inputs per gene are generically in the chaotic regime if Boolean functions are chosen randomly from the full range of possible functions of $K = 3$, $K = 4$, or more inputs. We have observed a substantial bias toward canalyzing functions. Does this bias suffice to tune networks with $K = 3$, or $K = 4$ or more genes into the ordered regime?

Our group constructed large networks of genes, using Wuensche's wonderful DDlab program, available on line, to examine model systems with up to 65,000 genes. When we made networks with $K = 3$ or $K = 4$ inputs and randomly chosen

Boolean functions, their Derrida curves, as expected, were in the chaotic regime, a percolating green sea existed, and vast purple avalanches careened around the system.

When we made networks with $K = 3$ or $K = 4$ inputs, tuned to the exact distribution of fractions of genes with 0, 1, 2, 3, or 4 canalyzing inputs, the results (Figure 8.9a) show that such networks are clearly in the ordered regime. The Derrida curve is below the main diagonal. Therefore, in such networks a percolating frozen red sea exists, leaving behind isolated green islands, and the distribution of purple damage avalanches is a power law with a finite cutoff at about 2 times the square root of the number of genes (Figure 8.9b). This last predicts that the largest avalanches of gene changes if any single gene is perturbed in humans should be about 634 genes. This fits presently known data.

We even have tentative evidence of detailed evolutionary tuning. As the number of inputs per gene increases, a gradually decreasing fraction of the Boolean functions must be canalyzing to cross the phase transition into the ordered regime. Although the data are too few to warrant conclusion, the fraction of canalyzing inputs for $K = 3$, $K = 4$, and $K = 5$ eukaryotic genes trends downward as K increases along the curve needed to remain just within the ordered regime. The observed fraction of canalyzing functions are: 0.795, 0.708, 0.649 for $K = 3$, 4, and 5, respectively. If so, only natural selection can have tuned it thus.

There are other clues, reported in *Origins of Order* and *At Home in the Universe* in some detail, that support the hypothesis that cells lie in the ordered regime. This interpretation is based on the assumption that the different cell types of a higher eukaryote correspond to the different state cycle attractors of the network. One attractor is a liver cell, another is a kidney cell, and so forth.

- The percolating frozen core that is identical on all attractors of the Boolean network is likely to correspond to the core set of genes whose expression is known to be identical on all cell types, commonly thought to be housekeeping genes.
- The typical differences in gene activity patterns in model cell type attractors, usually a few percent, mirror the data for real cells.
- The number of state cycle attractors robustly scales as a square root function of the number of genes in the ordered regime. The number of cell types in real cells scales as roughly a square root to a linear function of the estimated number of genes in that organism, from yeast to sponge to worm to man. Indeed, the square root of 80,000 is about 273, and Bruce Alberts and colleagues quote the number of cell types in humans as 265.
- The expected power law size distribution of avalanches of gene changes after perturbation of a single gene's activity seems plausible and fits the still sparse data.

Derrida Plots, Data vs. Random

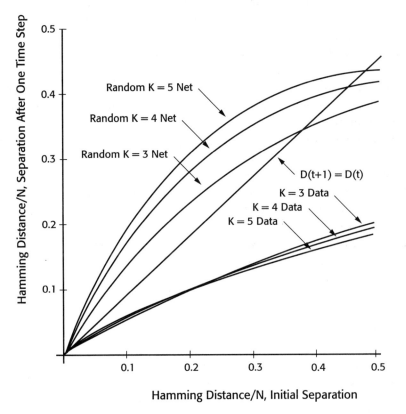

FIGURE 8.9a Comparison of the Derrida curve for model genetic networks with N = 1000 model genes and with K = 3, 4, or 5 inputs and randomly chosen Boolean functions, which all lie in the chaotic regime, and model genetic networks where the fraction of canalyzing functions, and numbers of canalyzing inputs per gene were matched to published data for real eukaryotic genes with K = 3, 4, or 5 known inputs. All three Derrida curves based on the observed bias toward high numbers of canalyzing inputs per gene are well into the ordered regime and virtually identical. This near identity suggests that natural selection is tuning the numbers of canalyzing inputs per gene for genes with different numbers of regulatory inputs to achieve the same location in the ordered regime (see Figure 8.5b). These results suggest strongly that eukaryotic cells are well inside the ordered regime.

- Model cell types are homeostatically stable to most small perturbations. So are real cell types.
- If a state cycle attractor is a cell type, then cellular differentiation from one to another cell type corresponds to a perturbation that causes the cell type to

leave one attractor and flow to another attractor. In the ordered regime, any cell type can only directly reach a few adjacent cell types and may, by a succession of perturbations, eventually differentiate along branching developmental pathways to a larger number of cell types. Precisely this pattern of branching differentiation is known in all multicelled organisms.

There are other data, but perhaps that will suffice. I believe the initial evidence strongly suggests that eukaryotic cells are in the ordered regime, not too far from the phase transition to chaos.

This hypothesis, which I here tentatively adopt as a candidate general law for any

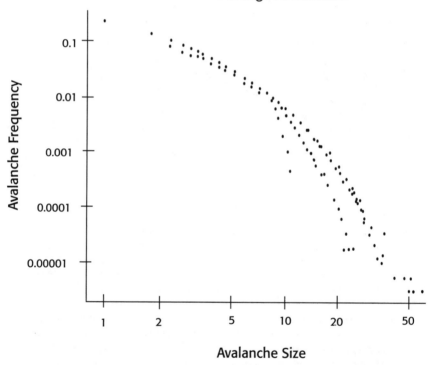

FIGURE 8.9b Power law distribution of sizes of avalanches of changes in gene activities for networks of three different sizes, N = 20, 80, and 320, and with a 50:50 mixture of K = 3 and K = 4 genes exhibiting the observed biases toward high numbers of canalyzing inputs per gene for eukaryotic genes. The numerical data represent a total of about 300,000 avalanches. The size of the largest avalanches scale as a function of the square root of the size of the genetic network.

biosphere—a very long jump to be sure—is now open to direct tests. Current technology, based on Affymetrix chips, displays the DNA from thousands of different genes in a two-dimensional array. RNA can be sampled from small tissue fragments, or even single cells, and, via a few steps, caused to bind through Watson–Crick base pairing to the corresponding DNA sequence. In this way, the transcribed RNA abundances of thousands of genes can be sampled simultaneously. Thus, we can now follow the RNA states of cells over time, in normal and diseased states, treated and nontreated states, and so forth. Companies such as Incyte are doing just this and selling the data to the large pharmaceutical companies for analysis.

But then we can clone controllable cis sites such as promoters into cells at one or more randomly chosen sites and study the effects of transiently perturbing the activities of one or a few genes. Is the Derrida curve below the main diagonal or not? Does a power law distribution of avalanches of change erupt or not? We can use the data to find the genes in the same isolated green islands, for avalanches should be confined to one island and overlap if started at different genes in the same island. More, patterns of gene activities that change will change in correlated ways for genes in the same green island, but not for genes in different green islands.

Remarkably, recent evidence suggests just such correlated patterns of gene activity changes. John Welsh has analyzed the transcription patterns of almost 2,000 different genes in a specific cell type, the human melanocyte, from newborn children, subjected to the eight possible different combinations of three distinct modes of perturbation. Welsh could, in principle, distinguish increases, decreases, or no change in the abundances of gene transcripts for his nearly 2,000 genes. Of these, 1,695 showed no detectable change, about 280 showed changes. Given eight treatment regimes, the control, and the seven other treatments consisting of all combinations of one or more of his three perturbations, in principle, there are three raised to the seventh power, or about 1,800, possible patterns of response. But, surprisingly, the 280 genes showed only 32 patterns. Already this is unexpected.

But the most interesting result is that, of the 32 patterns, 16 fall into eight mirror-symmetric pairs: Under some conditions, one set of genes increases in transcript abundance while a second set of genes decreases in transcript abundance. Under other of the seven perturbing conditions, the roles are reversed, and the first set of genes decreases in transcript abundance while the second set increases in transcript abundance. Welsh found eight such mirror symmetric pairs of sets of genes, suggesting at least eight different coordinated sets of genes, each coregulated, yet each buffered from the other sets of genes.

It may be that Welsh has found the first evidence of genes lying in eight different green islands, buffered from one another by the percolating red frozen structure. If the green islands exist, they are the paragraph structure of the genome. They are the midsize decision-taking subcircuits of the genome. For each such island, cut off from influence by other islands by the frozen red structure, has its own alternative attractors, two for this island, five for that island, seven for a third island. The total number of attractors for the entire network is then $2 \times 5 \times 7 = 70$. And if so, cell types are a kind of combinatorial code of the choices made by the different islands.

And yet more: My colleague Marc Ballivet, with a minor bit of help from me, has come up with a means to rapidly clone most or all cis sites from cells. If the thousands of cis sites can be cloned, each can be used to affinity purify the trans factors binding it. Other biologists are learning how to construct small genetic circuits. By our means or others, the medicine of the twenty-first century will learn to control the activities of genes in genetic networks, hence, control tissue regeneration and differentiation. We enter the "postgenomic" era.

I return to my candidate law and remark next that cells typically do not live alone; they live in communities of single-celled organisms or other simple multicellular organisms, or they live in tissues in highly complex multicellular organisms. Thus, any candidate law must be considered with respect to a community of autonomous agents.

Consider an ecosystem with different species of bacteria. Each species may secrete different chemical species, S, that impinge on a subset, C, of the other cell species. In Figure 8.10 I show a three-dimensional coordinate system. One axis shows the order-chaos axis for a single isolated bacterium, measured by Derrida curve criteria, with order on the left, near the origin, and chaos on the right. The remaining axes show C and S.

If a given cell is at the phase transition between order and chaos, and additional molecular inputs, S per cell, come from C different types of cells, the total connectivity in the cell is raised from KN to $KN + CS$. The results will typically drive that cell into the chaotic regime. Indeed, the entire community will be driven into the chaotic regime.

Hence, in the coordinate system of Figure 8.10, Igor Yakushin at Bios Group has shown that a hyperbolic surface, as shown, separates the ordered regime from the chaotic regime. Cells can buffer themselves from chemical perturbations from other species by retreating deeper into the ordered regime. This can be accomplished by increasing the number of canalyzing inputs per gene. And indeed, as described above, individual eukaryotic cells do appear to lie well within the ordered regime, perhaps as buffering for the fact that such cells, like yeast, live in microbial communities or, like human cells, live in tissues where each cell is bombarded with chemical signals from other cells in the same body.

It is a plausible conjecture that communities of cells, and tissues, come to lie on the phase transition surface between order and chaos. The hypothesis is readily tested. If so, perturbations of a single gene in one type of cell should trigger a power law distribution of avalanches of changes of gene activities that spreads from the perturbed gene to other genes in that cell and species, and to cells of other species. Further, the Derrida curve of the total community should be at the phase transition.

Phase Transition Surface for the Colony of Cells

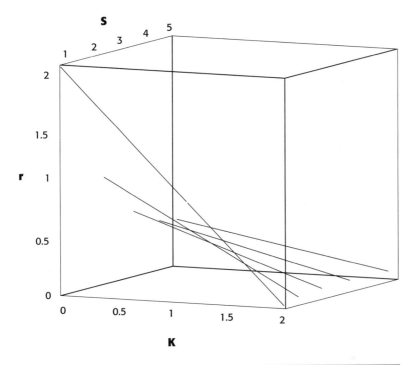

FIGURE 8.10 The generalization of the phase transition surface to colonies, or tissues, of cells, where each cell alone has K inputs per gene, but interacts with S other cells and exchanges a normalized fraction r = C/N of the gene product couplings between cells. In the absence of biases toward high canalyzing inputs per gene, or high P values, the phase transition occurs for single isolated cells at K = 2. As such cells are perturbed by interaction with r = C/N gene products from each of S other cells, the cells must retreat deeper into the ordered regime by decreasing K to avoid being driven into chaotic behavior by the perturbations each receives from the S other cells. The figure shows the phase transition surface, below which cells and the colony remain in the ordered regime, above which the cells and colony are in the chaotic regime. Real cells have many genes with more than K = 2 inputs and can retreat deeper into the ordered regime by increasing P, or the fraction of canalyzing inputs per cell. I conjecture that colonies and tissues arrange S and C/N and biases in favor of canalyzing inputs per gene such that the colony or tissue resides on the phase transition surface or very near it in the ordered regime. Here the maximum number of reliable discriminations can be made and acted upon without trembling hands. Since real cells have K > 2 inputs and appear to have a sufficient bias toward high numbers of canalyzing inputs per gene such that individual isolated cells are well within the ordered regime (Figure 8.9a), there is room for chemical couplings between cells to bring tissues or colonies close to the phase transition boundary to chaos.

A final numerical test is under way. The aim of the numerical test is to check whether cells and communities that lie at the phase transition would, in fact, make the maximum number of reliable discriminations and act without trembling hands.

Cells that are yammering at one another probably never reach their attractors. Pick a subset, M, of the states in each cell in the community as its "action" states. Now release the community, numerically, somewhere on the edge-of-chaos surface. Over a long period of time, the M action states will be encountered in some order, yielding some probability distribution of transitions between pairs of the M states, as each cell is perturbed by S chemical signals from C other cell types. The transitions among the M action states of each cell can be written in a matrix showing the transition probabilities between any pair of the M states. From this it is possible to calculate the mutual information, MI, between pairs of the M action states.

MI is $H(A) + H(B) - H(AB)$. Here $H(A)$ is the entropy of A, $H(B)$ the entropy of B, and $H(AB)$ is the joint entropy of A and B. If A and B are occurring randomly with respect to one another, then $H(AB)$ equals the sum of $H(A) + H(B)$, so MI is 0. If either A or B is unchanging, MI is again 0. But if A and B are changing in correlated ways, MI is positive. Hence, we can ask what value of M maximizes the mutual information among the M states, how is that related to cell and community position on the edge-of-chaos phase transition in Figure 8.10, and how well does the mutual information among the M action states correlate with the mutual information in the patterns of CS input signals arriving at each cell type? More broadly, can selection maximize both M and that mutual information correlation and, if so, for what value of M and where on or off the phase transition surface in Figure 8.10?

I do not know the answers but hope the optimal point lies on the phase transition surface, for such selected mutual information correlation would begin to show that such communities of cells with such regulatory networks can indeed make the maximum number of reliable discriminations and act on them without trembling hands to make a complex living in a complex world.

Candidate Law 2: Community Assembly Reaches a Self-Organized Critical State

First, a foray into self-organized criticality, a concept that will drive the rest of this chapter.

Per Bak and his colleagues in 1988 published a paper concerned with sand piles. One is to take a large, flat table, supply lots of sand, and gently let sand fall from on high onto the table. As the sand piles up, it eventually reaches the rest angle for sand and also extends to the boundaries of the table. You keep adding sand slowly. Sand-slide avalanches begin to form and sand drops to the floor. Measure the size distribution of the avalanches and a power law distribution is revealed, with many small avalanches and few large avalanches (Figure 8.11).

Power law distributions can, in fact, arise in many ways. One of those ways is at a phase transition, for example, in a ferromagnet at the phase transition temperature. Above the phase transition temperature, the magnetic spins line up randomly with one another and keep flipping. Below the phase transition

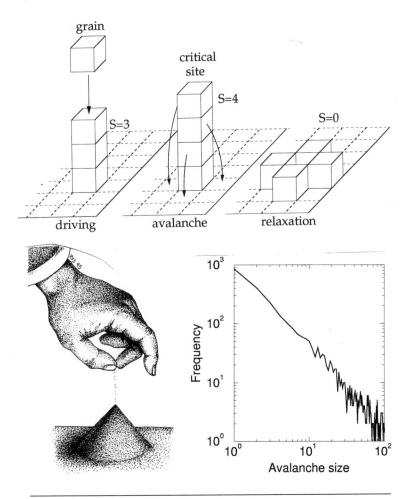

FIGURE 8.11 The sand pile model of Bak and colleagues. Idealized sand pile in which continuously added cubes of sand pile on top of one another. The maximum difference in height that can be sustained is S = 3 grains. As a next grain lands on the stack, yielding S = 4, the stack tumbles with one cube of sand falling in each of the four directions to yield the S = 0 configuration. Repeated addition of sand at random positions on a table top builds up the sand pile until the avalanche sizes reach a stable power law distribution, seen as a linear relation in the plot of the logarithm of frequency of avalanches of a given size against the logarithm of the size of the avalanche. This property of the system is called self-organized criticality.

temperature, the ferromagnet tends to line up with magnetic spins all pointing the same way, say, north pole upward, hence, the material is magnetized. At the phase transition temperature, something magic and "universal" occur: clusters of spins oriented the same way arise, and the clusters have a power law distribution of sizes. The power law distribution implies that there is no preferred size scale in the system. If there were a preferred size scale, clusters would be distributed exponentially in size, setting a size scale at which clusters at that size were half as likely as tiny clusters of spins.

So too for the sand pile. There is no preferred size scale revealed by the power law distribution of sand slide avalanche sizes. The big difference is that clever physicists tuned the temperature of the ferromagnet to the critical phase transition temperature. Per and company tuned nothing, they merely let sand drop randomly and gently onto the sand pile, and the sand pile tuned itself to criticality, hence the name: self-organized criticality.

Many of us have fallen in love with these results. Many are critical as well. I am on the pro side of the debate. I know the theory does not apply to real sand, with its rough edges, and works well with short-grain Swedish rice, but I love it anyway. Newton's law of universal gravitation does not work for bits of paper and cannon balls falling from the Tower of Pisa—wind resistance, you see. The bits of paper do not hit the ground at the same time as do the cannon balls. Most theories have ceterus paribus clauses. But I suspect Per and friends have found a deep truth about how nonequilibrium systems self-organize.

The remaining three of the four candidate laws apply versions of this idea. Even law 1 above is a version. Selection tunes communities of cells to the phase transition between order and chaos where a power law distribution of damage avalanches on all scales propagate across the system.

The claims coming next are not my own work, but derive from fine efforts by ecologists Stuart Pimm, Mack Post, and more recently, Bruce Sawhill and Tim Keitt, making use of work by physicist Scott Kirkpatrick and his colleagues.

First, the early work of Stuart Pimm and Mack Post, done in the late 1970s: They were concerned with community assembly of organisms into a local ecosystem (Figure 8.12). They ignored long-term coevolution and made use of the Lotka-Volterra equations. These equations basically say, for any species, what other species it eats, how readily it turns the eaten prey into an extra copy of itself, and how fast it reproduces on its own without eating. Plants, herbivores, and carnivores are readily represented.

Stuart and Mack did a surprising study with an astonishing and still poorly understood result. They made a pile of hypothetical critters in a computer, each governed by some plausible Lotka-Volterra equation whose parameters were drawn at random from some distribution. Then, with fond hopes, Stuart and Mack randomly chose species, one after another and tossed them into the computational

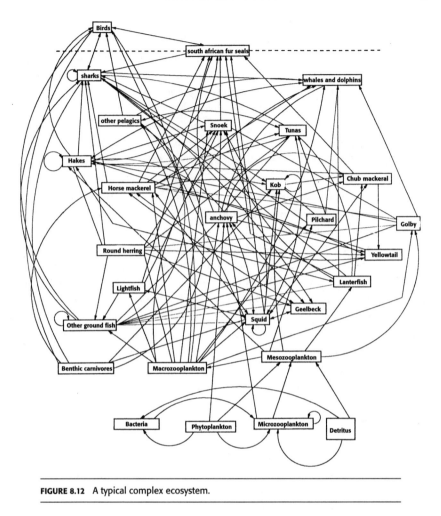

FIGURE 8.12 A typical complex ecosystem.

equivalent of east Kansas. ("Kansas is a place for people who like subtlety," my friend Wes Jackson once told a group of us visiting his Land Institute in Kansas in January. We observed subtly different shades of brown grass, trees, dirt, dust, hawks, mice, and agreed.)

What Stuart and Mack found was this: When the first few species were tossed into east Kansas in silico, they tested whether these few could coexist by running the corresponding Lotka-Volterra equations of all the species simultaneously to see if the hypothetical species all sustained abundances above zero. For example, the model community might go to a steady-state ratio of four species or might enter a limit cycle oscillation or some chaotic dynamics with a strange attractor that remains above zero for all species.

Dandy, the mock community was stable. They kept adding randomly chosen species. At first, it was easy to add new species, but it became progressively harder until no more species could be added. The deep mystery is, why? It is not a lack of food or energy. Furthermore, on the way to assembling the community, Stuart and Mack began to notice that as the community filled up, addition of one species could make one or more other species go locally extinct. They began to find evidence of power law distributions of such extinction events.

Why? No one knows for sure, and Stuart has some wonderful ideas discussed in his fine book, *The Balance of Nature*. But I am going to move on to the recent ideas of Bruce Sawhill and Tim Keitt, which build on the work of physicist Scott Kirkpatrick. Scott is famous for coinventing the Sherrington-Kirkpatrick spin-glass model and hangs out at IBM being smart. Not long ago, he took up what is called the "Ksat" problem.

Here is the Ksat problem. Consider some logical formula, or expression, as in our Boolean net. Any Boolean expression can be cast in "normal disjunctive form"; an example is (A_1 or A_2) and (A_3 or A_4) and (not A_1 or not A_4). In such an expression the variables between brackets constitute a "clause," so this logical expression has three clauses. The clauses are linked by "and." Thus, for the entire statement to be true, all three clauses must be true. Within each clause, variables are linked by the logical "or," symbolized with v. ($A_1 v A_2$) is true if A_1 is true, if A_2 is true, or if both A_1 and A_2 are true.

Now we can ask if the above expression can be satisfied by some assignment of true or false, 1 or 0, to the four variables, A_1, A_2, A_3, and A_4. The answer is yes since if A_2 = true and A_3 = true and A_4 = false, then all clauses are satisfied. On the other hand, consider (A_1) and (not A_1). There is no assignment of true or false to A_1 that makes both clauses true since they contradict one another. More generally, an expression in normal disjunctive form, with K variables ($A_1 v A_2 v \ldots Ak$) in each clause, a total of C clauses and a total of V variables $A_1, A_2, \ldots Av$ may or may not be satisfiable.

The wonderful result that Scott and friends showed is a phase transition from Ksat expressions that are almost certainly satisfiable to Ksat expressions that are almost certainly not satisfiable (Figure 8.13).

In Figure 8.13, the horizontal axis is labeled C/V. Thus, the x-axis shows the ratio of clauses to variables, hence, on average, how many clauses each variable is in. Obviously, as each variable is in more and more of the C clauses, with randomly assigned truth requirements, Vi versus not-Vi, the chance that the set of clauses can be jointly satisfied gets harder. On the other hand, as K goes up, there are more variables per clause, any one of which, if satisfied, satisfies the clause since the variables within a clause are joined by "or." Thus, as K goes up, the problem gets easier.

Remarkably, there is a phase transition on the C/V axis at $\ln 2 \times (2^K)$ or 0.6×2^K. As shown in Figure 8.13, for C/V values less than this phase transition value, the expression is almost certainly satisfiable. As C/V passes the 0.6 x 2^K critical value, the probability that the expression can be satisfied plunges to near zero.

Now, the idea that Bruce and Tim had was that building a community with random critters having random food and niche requirements is like the Ksat problem. They consider S species. Each species' niche includes some of the S species that it eats and some that eat or kill it. Thus, species $S7$ must eat $S3$ or $S5$ or $S9$ to survive, but can survive on its own only in the absence of $S10$, which poisons the chloroplasts that allow $S7$ to be an autotroph. In disjunctive form, the requirements for species $S7$ to survive are $(S3 \text{ v } S5 \text{ v } S9)$ and (not $S10$).

Bruce and Tim did numerical experiments for different values of S and K and C where again K is the number of alternative species any given species could eat. $(S3 \text{ v } S5 \text{ v } S9)$ corresponds to $K = 3$. They found the same phase transition. As more and more species are added, there are more potential interactions among the species since the pairwise possibilities increase as the square of the total species diversity. As this occurs, each species appears in an increasing number of clauses, so as C/V increases, at some point the satisfiability of the Ksat system went from easy to hard. It became hard, then impossible, to add new species. The community filled up because the Ksat problem went from easy to impossible.

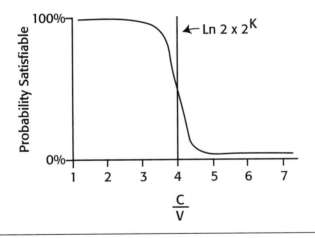

FIGURE 8.13 The phase transition in Ksat problem solvability (see text). As the ratio of clauses, C, to variables, V, increases, each variable in a randomly chosen Boolean expression in normal disjunctive form appears in more clauses. This increase renders it less likely that there is an assignment of true or false to each of the V variables that makes the entire Boolean expression true. As C/V increases, a sudden phase transition decrease in the probability of satisfying the Boolean expression occurs. The midpoint of this decrease occurs at a value that depends upon the number of variables, K, in each clause. The midpoint is at ln 2 x 2K.

I find this line of thinking rather interesting. Given rather general assumptions on the probability per pair of species of who eats or kills what, assigned more or less randomly, then as the number of species, hence pairs of species increase, such communities can fill up in the presence of persistent attempts at invasion. Bruce and Tim may have found the underlying reason for Stuart and Mack's earlier results. Moreover, Bruce and Tim have found reasonable numerical evidence for small and large avalanches of local extinction events upon entry of new species while the community was filling up.

Meanwhile, experimental work assembling communities of real organisms shows much the same results. Communities tend to fill up and do exhibit small and large local extinction events.

Candidate Law 3: Coevolutionary Tuning of Fitness Landscapes and Organisms to a Self-Organized Critical State

Begin with a well-stated claim of Darwin: gradualism. Species evolve, argued Darwin, by the gradual accumulation of useful variations that were gradually sifted by natural selection.

Darwin is correct about contemporary life, and presumably about ancient life, based on the record. In fact, for current life forms, seven decades of hard work by geneticists, working with organisms as disparate as mouse, fruit fly, maize, yeast, and many other eukaryotes, demonstrates conclusively that most mutations are of minor effect. For example, the fruit fly, *Drosophila melanogaster*, upon which I worked for twelve years, has the abdominal bristles alluded to above. A modest number of mutants exist that slightly increase or slightly decrease the number of abdominal bristles. The flies don't seem to mind, at least in the odd security of my and other biologists' laboratories.

More rarely, there are mutants of rather dramatic effect, none more so than the famous homeotic mutants of *Drosophila*, which fascinated me and many others. Here a single mutant can change an antenna into a leg or an eye to a wing or a head to the genitalia. These survive perfectly well in the laboratory as well, but one expects would not fare well in the real world.

Therefore, most of the heritable variation cast up by mutation is of minor effect, and gradual variation is persistent grist for the selection mill. Somehow, current organisms have contrived themselves to be such that most mutations are of minor effect. But is it necessarily the case that all complex systems have the property that most mutations are of minor effect? And if not, where does the gradualism of Darwinian selection come from?

Importantly, it is easy to create systems that are not readily adaptable by mutation and selection. What follows is not quite a theorem (and was discussed in *At Home in the Universe*), but it will do. Many of the readers of this book are competent programmers. Consider a typical program, say written in C, Java, or some other language. Perhaps it computes something as simple as the square roots of the first two million integers. Perhaps it simulates an ecosystem. Whatever that program does, imagine trying to evolve it by making random mutations in the code.

We all know what would happen. Most mutations of a computer program are of major effect. The program won't compile. If it compiles, it generates some vast stream of symbolic nonsense or goes into an undetected infinite loop and "hangs."

And we can make the matter substantially worse by eliminating redundancy in the code. Any computer program can be written as a sequence of binary, 1 and 0, symbols, where that sequence represents the input data to the program and the program itself.

Now, a well-known area of computer science considers how redundant a program is; for example, a simple redundancy would duplicate each binary symbol. Computer scientists talk of eliminating the redundancy of a computer code to achieve the most compressed possible code. A fascinating theorem states that there is no proof that a given computer program is maximally compressed, but that if it is maximally compressed, it is in a rigorous sense not detectably different from a random sequence of binary digits.

Figure 8.14 shows a four-dimensional Boolean hypercube with all two to the fourth, or sixteen, possible binary sequences of length four, ranging from (0000) to (1111). Each sequence is on one of the sixteen vertices of the hypercube and connected to four 1-mutant neighbors achieved by changing a single binary symbol among the four from 1 to 0 or from 0 to 1. Imagine that the minimal program we were considering were a 1000-long binary sequence. Then that minimal program could be represented as a single vertex on the 1000-dimensional Boolean hypercube with 2 to the 1000th = 10 to the 300th different sequences, hence vertices. A remarkable theorem due to Gregory Chaiten shows that if there is a minimal program with 1000 binary symbols, there is at most about one such minimal 1000-bit symbol sequences. In short, only a single vertex on the 1000-dimensional Boolean hypercube corresponds to the desired minimal program. Now consider each of the other 1000-bit sequences on the hypercube as a computer program. (Consider a gedankenexperiment, a thought experiment, for I have not carried out the actual computer experiment, and no one has been able to prove my following plausible, probably true, conjecture.) Run each binary sequence as the input data and program on our universal computer. Measure in some sense how far away the printout of that program is from the correct program for some finite chunk of the correct program's printout. If one started with the correct program, my bet is that since all redundancy has been removed, any single mutation to the code will randomize what the code does. Distant 1000-bit binary strings will, on average, be as good or bad an approximation of the correct program as its 1000 1-mutant variants.

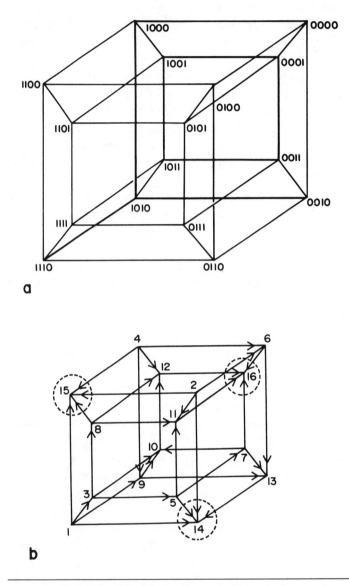

FIGURE 8.14 The four-dimensional Boolean hypercube. Each vertex has an assignment of 1 and 0, values for each of the four variables. Each vertex is next to four neighbors that each differ in the value, 1 or 0, assigned to one of the four variables. Any computer program can be written as a binary string of some length, N, hence is located as one vertex of the N-dimensional Boolean hypercube.

If one thinks of the measure of how close the output of a binary string program is to the correct program as the "fitness" of that trial binary string, then the fitness can be thought of as a height. The distribution of heights over the 1000-dimensional Boolean hypercube therefore creates a fitness landscape. In fact, my conjecture amounts to stating that the resulting fitness landscape is completely random. Neighboring points have fitnesses that have no correlation.

Assume my conjecture is true. There are theorems stating that there is no way to hill climb to the global peak, the single correct program, by accumulating mutants that gradually improve the program. Indeed, the only way to find the single good program is, effectively, to search the entire 1000-dimensional Boolean hypercube. You'd have to look at most of the 10^{300} vertices on that cube to find the working program. The problem is, as they say, *NP* hard, meaning the size of the problem scales exponentially in the length of the binary symbol sequence. But a single example makes a general point: Not all complex systems can be assembled by an evolutionary process!

It follows that only some complex systems can be assembled by an evolutionary process. And it turns out that an evolutionary process based on mutation, recombination, and selection, the genetic search mechanisms of current life, does very well on a special kind of fitness landscape, where the high peaks tend to cluster near one another and the sides of the peaks are reasonably smooth, rather like the high Alps.

Then, as I first asked in chapter 1, where do such correlated fitness landscapes come from? More generally, I recall the no-free-lunch theorem of Bill Macready and David Wolpert. Macready and Wolpert wondered whether, averaged over all possible fitness landscapes, some search algorithms, such as mutation and selection, on average outperform all other search algorithms, such as random search on the landscape or hill descending or picking birthdays, taking their square, and jumping that distance in a randomly chosen direction.

The no-free-lunch theorem proves that, averaged over all landscapes, no search algorithm outperforms any other. Well, my goodness! On average, random search and hill descending do just as well as hill climbing in finding peaks of high fitness. And here we organisms are, stuck using mutation, recombination, and selection. Yet organisms and ecosystems seem to be pretty complex. Once again, where did the "good" landscapes come from, the ones that Darwinian gradualism works so well in searching?

In chapter 3 I was led by the above to define natural games as ways of making a living. Naturally, ways of making a living have evolved as organisms have evolved. So rather easily one gets to the conclusion that the winning games are the games that winners play.

And, as noted in chapter 3, those ways of making a living that set problems that are well searched out by the search mechanisms of organisms—mutation, recombination, and selection—will be well searched out. Many sibling species will arise and lineages will branch. There will be many species making livings and many cases of livings being made that are well searched out by the search mechanisms of organisms.

In short, there must be a self-consistent coconstruction of a biosphere in which organisms, ways of making a living, and search mechanisms jointly and self-consistently come into existence. Organisms are not solving arbitrary problems. We are solving the kinds of problems we can solve given our solution procedures. How could it be otherwise?

So, somehow—and we will have to seek plausible mechanisms—organisms are tuning the statistical structure of the fitness landscapes they are searching in evolution. But the problem is very much more complex than merely searching a fixed fitness landscape. Fitness landscapes are not fixed. If the abiotic environment changes, the fitness landscape of organisms changes, buckles, and deforms.

Worse, organisms coevolve. My favorite example remains the frog and the fly. If the frog develops a sticky tongue, the fitness of the fly is altered. But so too is the fitness landscape of the fly, what it should do next. It should develop slippery feet, or sticky stuff dissolver or a better sense of smell to smell sticky stuff before the frog gets too close or . . .

So, due to coevolution, the fitness landscape of each species heaves and deforms as other species make their adaptive moves.

A Sojourn to Coevolution in the NK Model

These results were presented in *At Home in the Universe* but are needed here. Since they are publicly available, I will be brief.

The *NK* model is a simple toy world in which an organism has N genes. Each gene comes in two "alleles," or versions, 0 or 1. Each allele of each gene makes a contribution to the fitness of the organism that depends on the allele of that gene and upon the alleles of K other genes. In genetics, these K other genes are called "epistatic" inputs to the fitness contribution of a given gene. The 2 to the N combinations of alleles of the N genes are therefore located on the vertices of the N-dimensional hypercube, like Figure 8.14. The fitness of each type of organism, or vertex, is written on that vertex and can be thought of as a height. Hence, the *NK* model creates a fitness landscape over the N-dimensional Boolean hypercube. To keep matters simple, I assume all critters have a single chromosome, that is, are haploids. When they are not feeling sexy, bacteria will do as an example.

Having chosen N and K, say $N = 3$ and $K = 2$, the rest is done at random in the hopes that generic features of N and K will show up in the resulting statistical structure of the fitness landscape (Figure 8.15a–c). In one limiting case, the K inputs to each of the N genes are chosen at random from among the N. Each gene has two alleles, 1 and 0. The fitness contribution of that gene is affected by which of its alleles occurs and by which of the 1 or 0 alleles of K other epistatic input genes occurs. Thus, each gene's fitness contribution is affected by $K + 1$ genes.

To study the generic features of such systems, I assign, once and for all, a random "fitness contribution" to each of the 2 to the $(K + 1)$ combinations of allele states affecting each of the N genes. The fitness contribution is drawn from the uniform interval between 0.0 and 1.0. Thus, instead of the Boolean functions described above showing when genes turn on and off, here I obtain a column vector for each of the 2 to the $(K + 1)$ allele states affecting a given gene, and in each position is a random decimal. Once this is done for each of the N genes, it remains to define the fitness of an organism with a specific allele at each of the N genes. I define this as the average of the fitness contributions of the N genes. The results yield a fitness landscape over the N-dimensional hypercube (Figure 8.15c).

I will briefly summarize results for the structure of NK landscapes. When $K = 0$, each of the N sites is independent. There is an optimal allele at each site and hence a globally optimal genotype. Any other genotype is suboptimal, but can steadily climb to the peak by flipping any gene in a less favorable allele to the opposite,

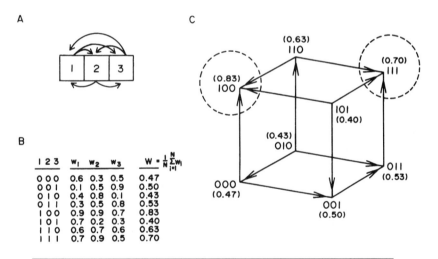

FIGURE 8.15 Building a fitness landscape. **(a)** The *NK* model of a genomic network consisting of three genes (N = 3), each of which can be in one of two states, or alleles, 1 or 0. Each gene receives inputs from two other genes (K = 2) that affect the fitness contribution of the gene. **(b)** Each gene in each of the 2³ = 8 possible genomes is randomly assigned a fitness contribution between 0.0 and 1.0. Then the fitness of each genome is computed as the mean value of the fitness contributions of the three genes. **(c)** A fitness landscape is constructed. Circled vertices represent local optima. Arrows on edges represent directions "uphill." An adaptive walk follows arrows, tail to head, from an initial position on the landscape upward to a local peak.

more favorable state, 0 or 1. So the landscape is like Fujiyama, single peaked with smooth sides.

When K is the maximum value, $N - 1$, as in Figure 8.15a–c, then each gene influences the fitness contribution of every gene. This is the totally interconnected system. Since fitness values are assigned at random for the 2 to the $(K + 1)$, or 2 to the N input configurations when $K = N - 1$, it is easy to show that the resulting fitness landscape is fully random.

A main feature of random landscapes is that there are nearly exponentially many local peaks, indeed the number of local peaks is 2 to $N/(N + 1)$. For $N = 1000$, there are 10^{297} local peaks on the landscape. Finding the global peak by hill climbing is improbable, and the system becomes trapped on a local peak. Other features include the lengths of walks via fitter neighbors to nearby peaks, which scales as the logarithm of N, and the way directions uphill dwindle on walks uphill. At each step uphill, the fraction of directions uphill is cut in half, yielding exponential slowing in the rate of finding fitter variants, hence, rather general laws about the rate of improvement slowing exponentially that we will discuss in the next chapter on "learning curves" in economics.

Now, on to coevolution and the evolution of the structure of fitness landscapes. Figure 8.16 shows a frog and fly, each characterized by an NK landscape, coupled

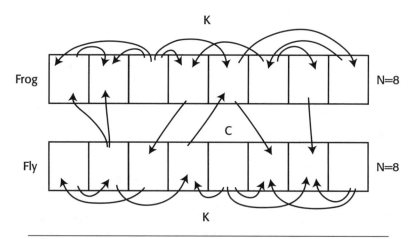

FIGURE 8.16 The frog and the fly in the context of the NK landscape model. Frog and fly each have N = 8 genes, or traits. Each gene or trait makes a contribution to the fitness of the frog or the fly that depends upon K other traits in the frog, or fly, respectively. In addition, genes or traits in the frog make fitness contributions that depend upon genes or traits of the fly, and vice versa, through the C couplings between frog and fly. Via these couplings between frog and fly, when the frog population moves on the frog's fitness landscape, that motion deforms the fly's fitness landscape, and vice versa.

together. Each of the N genes in the frog receives inputs from K genes in the frog and C genes in the fly, and vice versa. Thus, the sticky tongue of the frog affects the fitness of the fly via the presence or absence in the fly of slippery feet, sticky stuff dissolver, or a strong sense of smell for sticky frog tongues. To accommodate the C couplings, each gene in the frog looks at $K + C$ inputs and has its table of random fitness contributions augmented with new random decimals. So too for the fly feeling the effects of the frog.

Now, when the frog population moves by mutation and selection uphill on the frog landscape, those moves distort the fly's landscape, and vice versa. Coevolution is a game of coupled deforming landscapes. Figure 8.17a–c show coevolution in model ecosystems with four, eight, and sixteen species. Due to landscape deformations as species coevolve, an adaptive move by one species can cause the fitness of other species to decrease. In general, such coevolving systems can behave in two regimes, an ordered regime and a chaotic regime, separated by a phase transition.

In Figure 8.17a the four-species ecosystem eventually settles down to a state where fitnesses stop changing. This corresponds to an ordered regime or unchanging evolutionary stable state in which each species has evolved to a local peak on its fitness landscape that is consistent with the peaks occupied by its ecosystem neighbors. Once attained, each species is better off not changing so long as its neighbors do not change. By contrast, in the eight- and sixteen-species ecosystems, Figures 8.17b and c, fitnesses continue to jostle up and down as species evolve in a chaotic regime, each species chasing the adaptive peaks on its landscape that retreat—due to adaptive moves of other species—faster than each species can attain the peaks on its own landscape.

A major point of the coevolving NK landscape model is that the creatures can tune the structure of their fitness landscapes, each for its own selfish advantage. Yet, as if by an invisible hand, the tuned landscape structure works for the average benefit of all. This toy model is, to date, the only example I know in which creatures tune the structure of their fitness landscapes such that all evolve in problem spaces where, in some sense, they can search those spaces well self-consistently with their search mechanisms.

Here is how the toy model works. Each species is represented by a single individual. Hence, the species is assumed to be isogenic, except during the rapid evolution to fitter genotypes that happens as a fitter mutant of a species steps from one point to another point on the landscape. Very rapidly, it outreproduces its less-fit cousins. Hence, in this limit, the entire species can be said to hop between points on the landscape.

At each move of the computer program, any of four events may happen. A given species is chosen at random. First, it may do nothing. Second, it may change its genotype, and if the result is that it is fitter in interacting with its ecosystem neighbors, that innovation will be accepted. The creature has evolved on its landscape and prob-

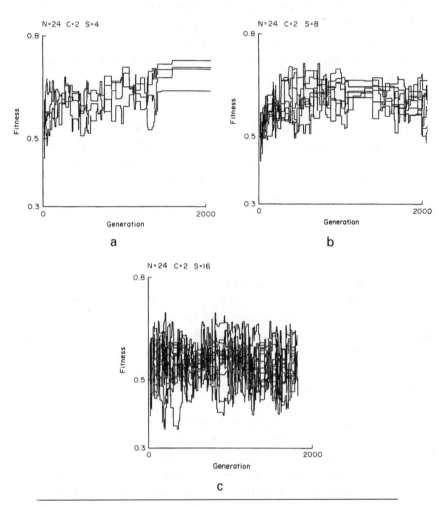

FIGURE 8.17 Order and chaos in the coevolutionary behavior of coupled species in model ecosystems with (a) S = 4, (b) S = 8, and (c) S = 16 species. When the system has four species, all species coevolve to fitness peaks that are mutually consistent and the system stops changing in a stable equilibrium. The coevolutionary system is in an ordered regime. As the number of species increases, mean fitness decreases and fluctuations in fitness increase. No coevolutionary stable equilibrium was found in 8,000 generations. The systems remained in chaotic "Red Queen" behavior.

ably deformed the landscapes of its neighbors. Third, the critter can change the ruggedness of its landscape by increasing K or decreasing K. Landscapes become more rugged and multipeaked as K increases. The move altering K is accepted only if that move makes the current genotype of the creature fitter. Hence, altering the ruggedness of the fitness landscape must pay off for the creature immediately and is accepted selfishly. The fourth thing that can happen is rather mean. A random other creature, say, Godzilla, is chosen to attempt to invade the current species niche. A copy of Godzilla, Godzilla', connects to the first species' ecosystem neighbors and has a go. If Godzilla' is fitter when coupled to the first species' econeighbors than that species, that species goes extinct in its niche and is replaced by Godzilla'.

By the fourth mechanism, if Godzilla' happens to have a beneficial landscape ruggedness due to its K value, that good landscape ruggedness has now replicated from the initial Godzilla to its copy. So good landscape ruggedness can spread, by natural selection, through the model ecosystem, hence, landscape ruggedness can evolve.

The results are shown in Figure 8.18. Indeed, landscape ruggedness does evolve to an intermediate ruggedness. During this evolution, the mean interval between

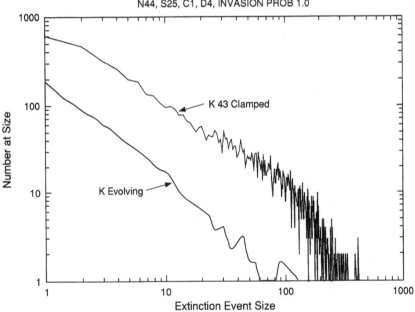

FIGURE 8.18 Controlling extinction events via co-evolution. Using logarithmic scales, the sizes of hypothetical extinction events are plotted against the number of events of each size. The result is a power law distribution. In the upper distribution, an ecosystem is held artificially deep in the high K regime, K = 43. Fitness is low in this regime because of conflicting constraints, so huge avalanches of extinction events cascade through the system. The lower power law distribution corresponds to a system where landscape ruggedness has been allowed to evolve to an intermediate K range, K = 22. Fewer and smaller extinction avalanches occur, so mean lifetime of species has increased. Interestingly, data for real extinction events in the record over the past 650 million years strongly suggest a similar gradual slowing in the size and number of extinction events and increase in mean species lifetimes. Evolution may have tuned landscape ruggedness and couplings between landscapes as in the present model.

extinction events increases dramatically, hence, the mean number of extinction events decreases. In this sense, all the creatures that remain become fitter due to the coevolutionary tuning of landscape ruggedness. In addition, when Godzilla' replaces the hapless species that now goes extinct, its econeighbors find themselves interacting with Godzilla' itself. They may not be as fit interacting with Godzilla' as with the first species, hence, they too may be invaded and driven extinct in turn.

In short, avalanches of extinction events can propagate. Figure 8.18 shows that the distribution is a power law, with many small and few large extinction events. Moreover, once a new species comes into existence, say, Godzilla', it may not fare well in its new niche, hence, may go extinct soon. But if it lasts in its niche, it may be well adapted, hence, resistant to being driven extinct. The results (Figure 8.20a) are a power law distribution of species lifetimes.

Thus, this model shows an invisible hand in which natural selection, acting on individuals only, tunes landscape ruggedness. All players are, on average, fitter in the sense of surviving as species for much longer periods, yet the ecosystem appears self-organized critical with a power law distribution of extinction events. Species lifetime distributions are also power laws.

Does this model apply to the real world? There is now considerable evidence that over the past 600 million years the size distribution of extinction events in the record, in terms of the number of species going extinct per 7-million-year period,

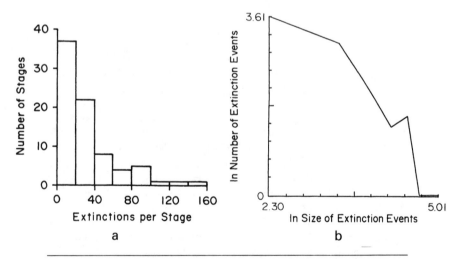

FIGURE 8.19 The real thing. David Raup has analyzed real extinction events according to the number of families that went extinct. (a) His data shows many small extinction events and few large ones. (b) When the data are replotted in natural logarithmic form, the result is not quite a linear power law distribution. More recent data fit a power law better.

is best understood as a power law, with many small and few large extinction events, (Figure 8.19). Furthermore, the lifetime distribution of species, as well as genera and families, is, indeed, a power law (Figure 8.20a,b).

I have discussed my own model primarily to focus on the fact that coevolution by self-natural selection alone acting on individuals alone can tune landscape ruggedness so those landscapes are self-consistently well searched by the creatures searching them and their search mechanisms. We can, and presumably do, self-consistently coconstruct ourselves, our niches (and hence problem spaces), and our search mechanisms such that, on average, we propagate ourselves and our descendants. It has worked for 4.8 billion years. Were our fitness landscapes such that we could not search them, we would not be here. We coconstruct our ways of making a living and search out better ways of making a living while we jiggle one another.

The *NK* model is but one crude model of coevolving organisms and their coupled deforming landscapes. More generally, each organism has traits that are

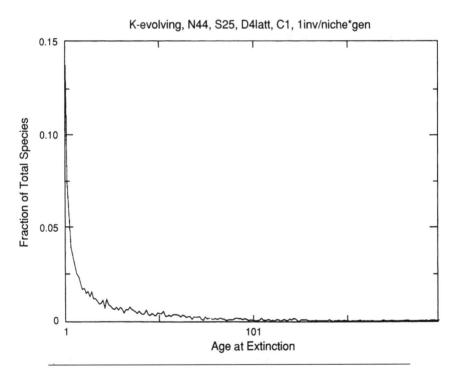

FIGURE 8.20a Infant mortality in artificial life. The age at which simulated species went extinct in the model by Kauffman and Neumann, plotted against the fraction of total species becoming extinct at each age. The distribution is similar to Figure 8.20b for real genera, both are power laws. Most species and genera die young, a few live a very long time.

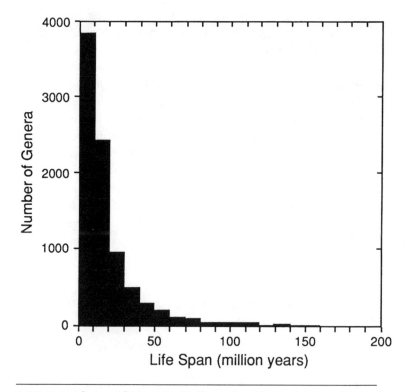

FIGURE 8.20b Infant mortality in marine life. The number of fossil vertebrate and inverte-
brate marine genera arranged according to their life spans. Most genera die young, but a
long tail extends to the right.

affected by many genes, the polygeny discussed above, and each gene affects many
traits, the pleiotropy alluded to above. It is interesting to note that were organisms
to evolve to a position below but near the biological reality that is the proper analogue of the Ksat phase transition, such a location might well achieve the gradualism and capacity to persistently evolve that Darwin noted and that we observe.
Both the gradualism and capacity to evolve are related to the number of alternative
assignments of true or false to the V variables that satisfy the Ksat normal disjunctive form. If there are connected pathways from one such assignment via 1-
Hamming-mutant neighboring assignments that all satisfy the normal disjunctive
form, then adaptive walks via alternatives genotypes are available, all of which
roughly generate the same organism. Gradualism is achieved. Polygeny and
pleiotropy tune landscape ruggedness and deformability, which tune coevolutionary dynamics, perhaps to a self-organized critical state of an ecosystem.
    There are, in short, dimly understood laws that allow the coevolutionary con-

struction of accumulating complexity. And it appears that such coevolution typically is self-organized critical. The *NK* coevolutionary model is not the only example of a model exhibiting self-organized critical behavior of model ecosystems. Bak and colleagues, Ricard Solé, and others have created elegant models aiming in the same direction. In particular, Solé's model comes closest to fitting the actual slopes of the observed power laws, which are -2.

Candidate Law 4: Expanding the Adjacent Possible in a Self-Organized Critical Way

How does the biosphere, collectively, broach and persistently invade the adjacent possible at the chemical, morphological, and behavioral levels?

I suspect there is a general law and mentioned it in the last section of the previous chapter. It is my hoped-for fourth law of thermodynamics for self-constructing biospheres. We enter the adjacent possible, hence expand the workspace of our biosphere, on average, as fast as we can.

Recall the Noah's Vessel experiment, with two of every species ground up in a blender, breaking all cell membranes, comingling the trillion or so proteins of the hundred million species with the thousands of small molecule metabolites. A supracritical explosion of chemical diversity would presumably ensue. As I noted, life has learned to avoid that fate. Cells are subcritical. Were they not, then any new chemical that chanced to enter the cells of Fredricka the fern would unleash a cascade of synthesis of novel molecular species, some of which would presumably kill poor Fredricka. Best defense? Stay subcritical. Why mess with that mess?

But recall that a mixed microbial community should be able to be driven to the subcritical-supracritical boundary by increasing the diversity of microbial species present and/or hitting the community with a sufficient diversity of novel small molecule species. My argument follows that if the community is supracritical, the novel cascading molecular species will kill off some of the microbial species, thereby lowering the community toward the subcritical regime. On the other hand, mutation and immigration should drive the community toward the supracritical regime. Do mixed microbial communities hover on the subcritical-supracritical boundary? We have seen other reasons that bound a community's complexity, including *K*sat problems noted in this chapter. So a better question is: Can microbial communities be driven supracritical? And if so, are they often near that boundary?

Of course, I do not know, but the hypothesis is testable. Increase the diversity of species and test for the diversity of synthesized small molecules, say, by gas chromatography. A colleague and I once devised such an experiment with mixtures of increasingly diverse moss species, planning to measure the molecular diversity of gas species evolved as a function of community diversity and the diversity in gas species introduced to the community. We went skiing instead. I still like the lines of the experiment. It could be carried out directly with mixed microbial communities as well.

As noted in the previous chapter, there must be some interplay in the entry into the adjacent possible that gates the exploration by the capacity of natural selection to trim away the losers. I described the Manfred Eigen–Peter Schuster error catastrophe. If the mutation rate in a population of viruses is low, by successive rare successful mutations the population climbs steadily uphill, then becomes trapped on or in the near vicinity of a local peak.

But let the mutation rate be increased. The population on the peak is deformed by the rapid accumulation of mutations and diffuses away from the peak into the lowlands of poor fitness. Tuning the mutation rate compared to the selection advantage of the fitness peak compared to nearby values on the fitness landscape tunes this error catastrophe. Above a critical ratio of the selective advantage at the peak to the mutation rate, the population remains near the peak. If the mutation rate is slightly higher, the population diffuses into the high-dimensional hinterlands, lost adrift in sequence space.

The error catastrophe is, of course, rather general. Eigen and Schuster consider a fixed high-dimensional sequence space, like our N-dimensional Boolean hypercube, which can be regarded as a sequence space in which molecules have only two "nucleotides" 0 and 1, C and G.

But what about an ever enlarging space of possibilities, expanding into an ever larger adjacent possible? Here too the rate of exploration of novel possibilities must be gradual enough that natural selection can weed out the losers. If not, the biosphere as a whole would diffuse into new ways of making a living so rapidly that selection would not be able to control the exploration, and we would soon falter.

So a general balance must be struck. We broach the adjacent possible by those exaptations that are not, I hold, finitely describable beforehand and do so at a rate that manages to work. We gate our entry into the adjacent possible.

Globally, for the entire biosphere, this suggests that we enter the adjacent possible about as fast as we can get away with it. On average, the global diversity of the biosphere has increased secularly. Indeed, it would be fascinating to know what has happened to microbial diversity in 4.8 billion years. There are bacteria eating rocks down there two miles below the surface, in hot thermal vents, in the cold of Antarctic frozen tundra and lake edges, all over the place.

I do not know the form of the law that governs this exploration. Perhaps it is locally self-organized critical in communities, multiplied by the number of effectively independent local communities in the biosphere. But I can make out a law in which the adjacent possible is invaded, such that diversity and coconstructed, coevolved complexity accumulate, on average, as fast as it can.

I can sense a fourth law of thermodynamics for self-constructing systems of autonomous agents. Biospheres enlarge their workspace, the diversity of what can happen next, the actual and adjacent possible, on average, as fast as they can. Clues include the fact, noted above, that for the adjacent possible of a $6N$-dimensional phase space to increase as the biosphere's trajectory travels among microstates, a secular increase in the symmetry splittings of microstate volumes must occur, such that different subvolumes go to different adjacent possible microstates. Eventually, such subvolumes hit the Heisenberg uncertainty limit. As I noted, organisms do touch that limit all over the place in the subtle distinctions that we make, turning genes on and off, smell sensors on and off, and eyes on and off, this way and that.

The whole of this chapter suggests that autonomous agents coevolve to be as capable as possible of making the most diverse discriminations and actions, take advantage of the most unexpected exaptations, coevolve as readily as possible to coconstruct the blossoming diversity that is, and remains, Darwin's "tangled bank." I sense a fourth law in which the workspace of the biosphere expands, on average, as fast as it can in this coconstructing biosphere.

A fourth law for any biosphere? I hope so.

Chapter 9

THE PERSISTENTLY INNOVATIVE ECONOSPHERE

I T IS NO ACCIDENT that the words for economics and ecology have the same Greek root, "house." Ecology and economics are, at root, the same. The economy of *Homo habilis* and *Homo erectus*, the stunning flaked flint tools of the Magdalinian culture of the magnificent Cro-Magnon in southern France 14,000 years ago when the large beasts had retreated southward from the glaciation, the invention and spread of writing in Mesopotamia, the Greek agora, and today's global economy are all in the deepest sense merely the carrying on of the more diversified forms of trade that had their origins with the first autonomous agents and their communities over four billion years ago.

Economics has its roots in agency and the emergence of advantages of trade among autonomous agents. The advantages of trade predate the human economy by essentially the entire history of life on this planet. Advantages of trade are found in the metabolic exchange of legume root nodule and fungi, sugar for fixed nitrogen carried in amino acids. Advantages of trade were found among the mixed microbial and algal communities along the littoral of the earth's oceans four billion years ago. The trading of the econosphere is an outgrowth of the trading of the biosphere.

Economics has considered itself the science of allocation of scarce resources. In doing so, it shortchanges its proper domain. Indeed, if we stand back and squint, it

is easy to see the most awesome feature of an economy and its roots in autonomous agents: The most awesome feature of the econosphere, as of the biosphere—both built by communities of autonomous agents in their urgent plunging,
lunging, sliding, gliding, hiding, trading, and providing—has been a blossoming
diversity of molecular and organismic species and of novel ways of making a living
that has persistently burgeoned into the adjacent possible. From tens of organic
molecular species to tens of trillions; from one or a few species of autonomous
agents to a standing diversity of some hundred million species and a total diversity
some hundred to thousandfold larger of those creatures come and gone.

Homo erectus had fire and early tools. *Homo habilis* traded stone axe parts 1.6
million years ago. The diversity of Cro-Magnon goods and services in the south of
France some 14,000 years ago may have numbered in the several hundreds to a few
thousands. Today, surf the web and count the diversity of goods and services, the
ways of making a living; it is in the millions.

Neither the biosphere nor the econosphere are merely about the distribution of
limited resources, both are expressions of the immense creativity of the universe,
and in particular, of autonomous agents as we exapt molecularly, morphologically,
and technologically in untold, unforetellable ways persistently into the adjacent
possible. Jobs and job holders jointly coevolve into existence in the econosphere in
an ever-expanding web of diverse complexity.

One of the most striking facts about current economic theory is that it has no
account of this persistent secular explosion of diversity of goods, services, and ways
of making a living. Strange, is it not, that we have no theory of these overwhelming
facts of the biosphere and econosphere? Strange, is it not, that we pay no attention
to one of the most profound features of the world right smack in front of our collective nose? And consistent with this strangeness, the most comprehensive theory,
the rock foundation of modern economics, the beautiful "competitive general
equilibrium" theory of Arrow and Debreu, does not and cannot discuss this explosion, perhaps the most important feature of the economy.

General Competitive Equilibrium and Its Limitations

So we begin with an outline of competitive general equilibrium as the cornerstone
conceptual framework of modern economics. Ken Arrow is a friend. As one of the
inventors of the framework, Ken is more at liberty to be a critic than the two generations of economists who have followed in his footsteps. My best reading of
Ken's view, which I share, is that competitive general equilibrium is, at present, the
only overarching framework we have to think about the economy as a whole. Yet
Ken suspects that that framework is incomplete; I agree.

Competitive general equilibrium grows out of a conceptual framework in
which the core question is how prices form such that markets clear. Recall the fa-

mous supply-and-demand curves for a single good (Figure 9.1). As a function of increasing price, plotted on the x-axis, supply, plotted on the y-axis, increases from low to high. More companies are willing to create widgets as the price per widget increases. On the other hand, demand, where demand is also plotted on the y-axis, decreases from high to low as prices increase. Fewer customers are willing to buy widgets as the price per widget increases.

As the figure shows, the supply-and-demand curves cross at some price. At that price, the markets "clear," that is, all the widgets supplied are purchased. The price at which markets clear is the "equilibrium price."

For a single good, the problem is simple. But consider bread and butter. Since many of us like butter on our bread, the demand for butter depends not only on the price and hence the supply of butter, but also on the price of bread, hence on the supply of bread. And vice versa, the demand for bread depends upon the price of butter and hence on the supply of butter. For thousands of goods, where the demand for any one good depends upon the price of many goods and the supply of any one good depends upon the price of many goods, it is not so obvious that there is a price for each good such that all markets clear.

But worse, the supply or demand for bread today may be different than the supply or demand for bread tomorrow. And still worse, the supply or demand for bread tomorrow may depend on all sorts of odd contingent facts. For example, if severe cold kills the winter wheat next month, the supply of bread will drop; if a

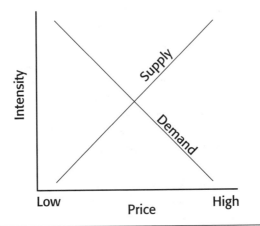

FIGURE 9.1 Textbook supply and demand curves for a single good, the supply curve increasing as a function of price, the demand curve decreasing as a function of price. Crossing point is the equilibrium price at which supply equals demand and markets clear. For an economy with many coupled goods, the requirements for market clearing are more complex (see text).

bumper crop of winter wheat comes available globally because of weather or suddenly improved irrigation and farming practices worldwide, the supply will go up.

Arrow and Debreu made brilliant steps. First, they consider a space of all possible dated and contingent goods. One of their examples of a "dated contingent
good" is "1 ton of wheat delivered in Chicago on May 2, 2008, under the condition
that the average rainfall in Nebraska for the six preceding months has been 10 percent less than normal for the past fifty years and that the Boston Red Sox won the
World Series the previous year."

In the Arrow-Debreu theory, we are to imagine an auctioneer, who at a defined
beginning of time, say, this morning, holds an auction covering all possible dated
contingent goods. All suppliers and customers gather at this imaginary auction,
bidding ensues for all possible dated contingent goods, with values calculated
under different hypotheses about the probabilities that the different dated contin-
gencies will come about. At the end of an imaginary hour of frantic bargaining, the
auction closes. All participants now have contracts to buy or sell all possible dated
contingent goods, each at a fixed price. Everybody hustles home to watch *Good
Morning America*. And, wondrously, however the future unfolds, whether there's
rain, sun, or snow in Nebraska, the dated contingent contracts that are appropriate
come due, the contracts are fulfilled at the preestablished price for each contract,
and all markets clear.

It is mind-boggling that Arrow and Debreu proved these results. The core
means of the solution depends upon what mathematicians call "fixed-point theorems." A beginning case is your hair, particularly for males, where short hair makes
the fixed point easy to see. When you comb your hair in a normal fashion, there is
a point roughly on top of your head, slightly to the back, where a roughly circular
swirl of hair occurs (ignoring baldness) around a fixed point where typically a bit
of scalp shows through.

A general theorem considers hair on a spherical surface and combing the hair
in any way you want. You cannot avoid a fixed point. More generally, replace hair
by arrows, with tails and heads, where each arrow is a line with an arrow head at
one end, drawn on the surface of the sphere. The arrows may bend if you like. Each
arrow can be thought of as mapping the point on the sphere at the tail of that
arrow to the point of the sphere at the tip of that arrow. So the arrows are a mapping of the surface of the sphere onto itself. For a continuous mapping, such that
there is a mapping from each point on the sphere, a general fixed-point theorem
proves that there must be at least one point on the surface of the sphere that maps
onto itself—that point is a fixed point under the arrow mapping.

The wonderful Arrow-Debreu general competitive equilibrium theorems depend on such a fixed point. In a space where all possible dated contingent goods
can be prestated and all possible markets for trade of all such goods exist, a fixed
point also exists that corresponds to a price at which all such markets clear, how-

ever the future unfolds. Arrow and Debreu won the Nobel Prize for their work, and won it deservedly. It is a beautiful theory. Yet there are important critiques of general competitive equilibrium. For example, the theorem depends on "complete markets," that is, markets to trade all possible dated contingent goods, and fine economists have raised issues about how well the theory works if markets are incomplete, as indeed they are.

I pose, however, a wider set of issues. The overarching feature of the economy over the past million years or so is the secular increase in the diversity of goods and services, from a dozen to a dozen million or more today. Nowhere does general competitive equilibrium speak about this. Nor can the theory speak about the growth in diversity of goods and services, for it assumes at the outset that one can finitely prestate all possible dated contingent goods. Then the theory uses complete markets and a fixed-point theorem to prove that a price exists such that markets clear.

But we have seen grounds to be deeply suspicious of the claim that we can finitely prestate all possible exaptations—whether they be new organic functionalities or new goods—that arise in a biosphere or an econosphere, such as Gertrude learning to fly in her terrified leap from the pine tree 60 million years ago last Friday or the engineers working on the tractor suddenly realizing that the engine block itself could serve as the chassis.

I do not believe for a moment that we can finitely prestate all possible goods and services. Indeed, economists intuitively know this. They distinguish between normal uncertainty and "Knightian uncertainty." Normal uncertainty is the kind we are familiar with in probability theory concerning flipping coins. I am unsure whether in 100 flips there will be 47 heads and 53 tails. Thanks to lots of work, I can now calculate the probability of any outcome in the finitely prestated space of possible outcomes.

Knightian uncertainty concerns those cases where we do not yet know the possible outcomes. Knightian uncertainty has rested in an epistemologically uncomfortable place in economics and elsewhere. Why? Because we have not realized that we cannot finitely prestate the configuration space of a biosphere or an econosphere; by contrast, Newton, Laplace, Boltzmann, Einstein, and perhaps Bohr have all more or less presupposed that we can finitely prestate the configuration space of any domain open to scientific enquiry. After all, as I have noted, we can and do prestate the $6N$ configuration space for a liter of N gas particles.

From this point of view, the wonderful Arrow-Debreu theory is fundamentally flawed.

Moreover, general competitive equilibrium, seen as a culmination of one central strand of economic theory, is too limited. Insofar as economics is concerned with understanding the establishment of prices at which markets clear, general competitive equilibrium was a masterpiece. But insofar as economics is or should be concerned with how and why economies increase the diversity of goods and services, the reigning theory is a nonstarter. And since the growth in wealth per capita over the past million years is deeply related to the growth in the diversity of technology and goods and services, contemporary economics is clearly inadequate.

We need a theory of the persistent coming into existence of new goods and services and extinction of old goods and services, rather like the persistent emergence of new species in an ecosystem and extinction of old species. In the previous chapter, we discussed ecosystems as self-organized critical. We discussed the biosphere and econosphere as advancing into the adjacent possible in self-organized critical small and large bursts of avalanches of speciation and extinction events. We discussed the power law distribution of extinction events in the biological record. And we discussed the power law distribution of lifetimes of species and genera.

But the econosphere has similar extinction and speciation events. Consider my favorite example: The introduction of the automobile drove the horse, as a mode of transport, extinct. With the horse went the barn, the buggy, the stable, the smithy, the saddlery, the Pony Express. With the car came paved roads, an oil and gas industry, motels, fast-food restaurants, and suburbia. The Austrian economist, Joseph Schumpeter, called these gales of creative destruction, where old goods die and new ones are born. One bets that Schumpeterian gales of creative destruction come in a power law distribution, with many small avalanches and few large ones. More, if species and genera have a power law distribution of lifetimes, what of firms? Firms do show a similar power law distribution of lifetimes. Most firms die young, some last a long time. We may bet that technologies show similar avalanches of speciation and extinction events and lifetime distributions.

The parallels are at least tantalizing, and probably more than that. While the mechanisms of heritable variation differ and the selection criteria differ, organisms in the biosphere and firms and individuals in the econosphere are busy trying to make a living and explore new ways of making a living. In both cases, the puzzling conditions for the evolutionary cocreation and coassembly of increasing diversity are present. The biosphere and econosphere are persistently transforming, persistently inventing, persistently dying, persistently getting on with it, and, on average, persistently diversifying. And into a framework of such a diversifying set of goods and services we must graft the central insights of general competitive equilibrium as the approximate short-timescale mechanism that achieves a rough-and-ready approximate clearing of markets at each stage of the evolution of the economy.

A rough hypothetical biological example may help understand market clearing in a more relaxed formal framework than general competitive equilibrium. Consider two bacterial species, red and blue. Suppose the red species secretes a red metabolite, at metabolic cost to itself, that aids the replication rate of the blue species. Conversely, suppose the blue species secretes a different blue metabolite, at metabolic cost to itself, that increases the replication rate of the red species. Then the conditions for a mutualism are possible. Roughly stated, if blue helps red more than it costs itself, and vice versa, a mixed community of blue and red bacteria may grow. How will it happen? And is there an optimal "exchange rate" of blue-secreted metabolite to red-secreted metabolite, where that exchange rate is the analogue of price?

Well, it can and does happen. Here is the gedankenexperiment: Imagine an ordered set of blue mutant bacteria that secrete different amounts of the blue metabolite that helps the red bacteria. Say the range of secretion is from 1 to 100 molecules per minute per blue bacterium, with metabolic cost to the blue bacteria proportional to the number of molecules secreted. Conversely, imagine mutant red bacteria that secrete from 1 to 100 of the red molecules valuable to the blue bacteria, at a similar cost proportional to the number of molecules secreted.

Now create a large, square petri plate with 100 rows and columns drawn on the plastic below to guide your experimental hands. Arrange the rows, numbered 1 to 100 to correspond to blue bacteria secreting 1 to 100 molecules a second. Arrange the columns, numbered 1 to 100 to correspond to the red bacteria secreting 1 to 100 molecules a second. Into each of the 100 x 100 cells on your square petri plate, place exactly one red and one blue bacterium with the corresponding secretion rates. Thus, in the upper left, bacteria that are low blue and low red secretors are coplated onto each square. On the lower left, high blue and low red secretors are coplated. On the upper right, low blue and high red secretors are coplated. And in the lower right corner, high red and high blue bacteria are coplated.

Go for lunch, and dinner, and come back the next day. In general, among the 10,000 coplated pairs of bacteria, while all 10,000 colonies will have grown, a single pair will have grown to the largest mixed red-blue bacterial colony. Say the largest mixed red-blue colony corresponds to red secreting 34 molecules per second, blue secreting 57 molecules per second.

This gedankenexperiment is important, for the exchange ratio of red and blue molecules is the analogue of price, the ratio of trading of oranges for apples. And there exists a ratio, 34 red molecules to 57 blue molecules per second, that maximizes the growth of the mixed red-blue bacterial colony. Since the fastest growing mixed red-blue colony will exponentially outgrow all others and dominate our gedankenexperiment petri plate, this red-blue pair establishes "price" in the system at 34 red to 57 blue molecules. Further, in the fastest growing red-blue colony, where red secretes 34 molecules and blue secretes 57 molecules per second, both the red and blue bacteria in that mixed colony are replicating at the identical optimum rate. As discussed in chapter 3, using a rough mapping of biology to economics, that rate of replication of a bacterium corresponds to economic utility and the increased the rate of replication corresponds to increased economic utility. The red and blue bacteria not only establish price, but they also share equally the advantages of trade present along the Pareto-efficient contract curve in the Edgeworth box discussed in chapter 3.

Mutualists in the biosphere have been hacking out rough price equilibria for millions of years and have done so without foresight and without the Arrow-Debreu fixed-point theorems. Indeed, critters have been hacking out rough price equilibria even as exaptations and new ways of living have come into existence and old ways have perished. Presumably, these rough biological price equilibria are reached because in the short and intermediate term they optimize the fitness of both of the mutualists. And the markets clear, in the sense that all the 34 red molecules are exchanged for 57 blue molecules per second. But it's a self-organized critical world out there, with small and large avalanches of speciation and extinction events in the biosphere and econosphere, and equilibrium price or no, most species and technologies, job holders and jobs, are no longer among us to mumble about advantages of trade.

I confess I am happier with this image of prices established in local, rough-and-ready ways at many points in an ecosystem or economy than with the beautiful fixed-point theorems of general competitive equilibrium. Bacteria and horseshoe crabs keep establishing rough price equilibria in their mutualisms without a prespecified space of ways of making a living. If they can do it, so can we mere humans. Getting on with it in the absence of predefined configuration spaces has been the persistent provenance of autonomous agents since we stumbled into existence.

Rational Expectations and Its Limitations

Actually, there has been a major extension of general competitive equilibrium called "rational expectations." Like general competitive equilibrium, this theory too is beautiful but, I think, deeply flawed.

Rational expectations grew, in part, out of an attempt to understand actual trading on stock exchanges. Under general competitive equilibrium, little trading should occur and stock prices should hover in the vicinity of their fundamental value, typically understood as the discounted present value of the future revenue stream from the stock. But, in fact, abundant trading does occur, and speculative bubbles and crashes occur. Rational expectations theory is built up around another fixed-point theorem. Rational expectations theory assumes a set of economic agents with beliefs about how the economy is working. The agents base their economic actions on those beliefs. A fixed point can exist under which the actions of the agents, given their beliefs about the economy, exactly create the expected economic behavior. So, under rational expectations one can understand bubbles. It is rational to believe that prices are going above fundamental value and thus to invest, and the investments sustain the bubble for a period of time.

Meanwhile, *Homo economicus* has been thought to be infinitely rational. In the

Arrow-Debreu setting, such infinitely rational agents bargain and achieve the best equilibrium price for each dated contingent good. In rational expectations, the agents figure out how the economy is working and behave in such a way that the expected economic system is the one that arises. The theories and actions of the agents self-consistently create an economy fitting the theories under which the agents operate.

But beautiful as these fixed-point theorems are, there are two troubles in the rational expectations framework. First, the beautiful fixed points may not be stable to minor fluctuations in agent behaviors. Under fluctuations, the economic system may progressively veer away from the fixed point into a feared conceptual no-man's-land. Second, achieving the fixed points seems to demand excessive rationality to fit real human agents. So it appears necessary to extend rational expectations.

One direction was broached thirty years ago, when economist Herb Simon introduced the terms "satisficing," and "bounded rationality." Both seem sensible but have been problematic. Satisficing suggests that agents do not optimize but do well enough; yet it has been hard to make this concept pay off. It has also been hard to make advances with the concept of bounded rationality for the simple reason that there is one way, typically, of being infinitely smart and indefinitely many ways of being rather stupid. What determines the patterns of bounded stupidity? How should economic theory proceed?

Natural Rationality Is Bounded

I suspect that there may be a natural extension to rational expectations applicable to human and any strategic agents, and I report a body of work suggested by me but largely carried out by Vince Darley at Harvard for his doctoral thesis. Two virtues of our efforts are to find a natural bound to infinite rationality and a natural sense of satisficing.

The core ideas stated for human agents are these: Suppose you have a sequence of events, say, the price of corn by month, and want to predict next month's price of corn. Suppose you have data for twenty months. Now, Fourier invented his famous decomposition, which states that any wiggly line on a blackboard can be approximated with arbitrary accuracy by a weighted sum of sine and cosine waves of different wavelengths and phase offsets, chosen out of the infinite number of possible sine and cosine functions with all possible wavelengths.

Now, you could try to "fit" the data on the corn prices with the first Fourier "mode," namely the average price. But presumably if the twenty prices vary a fair amount, that average will not predict the twenty-first month's price very well. You have "underfit the data." Or you could use twenty or more Fourier modes, all different wavelengths, with different phase offsets, and you would, roughly, wind

up drawing a straight line between the adjacent pairs of points in the twenty-pe-
riod series. This procedure will not help too much in predicting the twenty-first
period. You have "overfit" the data by using too many Fourier modes.

Typically, optimal prediction of the twenty-first period price will be achieved by
using two to five Fourier modes, each of different wavelength and different phase
offset. As is well known in the art, you have neither underfit nor overfit your data.

This fact suggests that the optimal prediction of a short sequence of data is obtained by a model of intermediate complexity—a few Fourier modes, neither a single one nor very many. The sense of bounded rationality Vince and I want to
advocate is that optimal prediction of a limited time series is achieved with models
using only a few Fourier modes, or their analogs in other basis sets—models of
modest, or bounded, complexity.

The rest of the theory Vince and I have developed goes to show that agents who
have theories of one another and act selfishly based on those theories will typically
create a persistently changing pattern of actions. Therefore, they persistently create
a nonstationary world in which only the relatively recent past has valid data. Thus,
there is always only a limited amount of valid data on which to base theories, and
the agents, in turn, must always build models of intermediate, bounded complexity to avoid over- or underfitting the meager valid data.

Natural rationality is, in this sense, bounded. It is bounded because we mutually create nonstationary worlds. What happens is that the agents act under their
theories. But in due course some agent acts in a way that falsifies the theories of
one or more other agents. These agents either are stubborn or change their theories. If they change their theories of the first agent, then typically they also change
their actions. In turn, those changes disconfirm the theory of the first agent, and
perhaps still other agents. So the agents wind up in a space of coevolving theories
and actions with no fixed-point, stable steady states, which means that past actions
are a poor guide to future actions by an agent since his theories, and hence his action plans, have changed. But this means that the agents mutually create a "nonstationary" time series of actions (nonstationary just means that the statistical
characteristics of the time series keep changing because the agents keep changing
their theories and actions). In turn, the agents typically have only a modest
amount of relatively recent data that is still valid and reliable on which to base their
next theories of one another. Given only a modest amount of valid and reliable
data, the agents must avoid overfitting or underfitting that smallish amount of
data, so they must use theories of intermediate complexity—for example, four
Fourier modes to fit the data, not one or twenty.

Vince and I want to say that natural rationality is bounded to models of intermediate complexity because we collectively and persistently create nonstationary
worlds together. In the agent-based computer models Vince has created for his thesis, just this behavior is seen. Indeed, we allow agents to evolve how much of the

past history of the interactions they will pay attention to and how complex their models of one another will be—one, four, or fifty Fourier modes. Agents evolve in a history and complexity space to find currently optimal amounts of history and complexity to use to optimally predict their neighbors. In our little world, the agents evolve to use a modest history, ignoring the distant past, and only modestly complex theories of one another.

We have found evidence of a further, perhaps generic, property that appears to drive such systems to settle down, then change in a sudden burst. As the system of agents and actions settles down to some repeatable behavior, an increasingly wide range of alternative theories, simple and very complex, fit the same data. But the complex theories, with many Fourier modes, attempt to predict fine details of the repeatable behavior. As those theories become more complex, they are more fragile because they can be disconfirmed by ever more minor fluctuations in the repeat-able behavior. Sooner or later such a fluctuation happens, and the agents with the complex disconfirmed theories change theories and actions radically, setting up a vast avalanche of changes of theories and actions that sweeps the system, driving the collective behavior far from any repeatable pattern. In these new circum-stances, only a small subset of theories fits the current facts, so the diversity, and complexity, of theories in the population of agents plummets, and the system finds its way back to some repeatable pattern of behavior.

In short, there appears to be not only a bounded complexity in our rationality, but a fragility-stability cyclic oscillation in our joint theories and actions as well. In these terms, the system of agents and theories never settles down to a fixed-point equilibrium in which markets clear. Instead, the system repeatedly fluctuates away from the contract curve then returns to new points in the vicinity of the contract curve. Hence, in precisely the sense of repeatedly fluctuating away from a contract curve then returning to its vicinity, the system does not achieve an optimizing price equilibrium, but satisfices.

The bounded complexity issues would seem to apply to any coevolving au-tonomous agents that are able to make theories of one another and base actions on those theories. The tiger chasing the gazelle and the starfish predating the trilobite are, we suppose, Popperian creatures able to formulate hypotheses about their worlds that may sometimes die in their stead. Presumably all such autonomous agents, under persistent mutation and selection, would opt for changeable models of one another of bounded complexity.

While these steps are only a beginning to go beyond rational expectations in economics, they seem promising. Whatever natural, or unnatural, games au-tonomous agents are playing as they and we coevolve in a biosphere or econo-sphere, nonstationarity arises on many levels. Here we see it at the level of the agents' theories of one another and the actions based on those theories. Perhaps this is just part of how the world works. Given the semantic import of yuck and yum, and the reality of natural games for fox and hare, for *E. coli* and paramecium, these changing theories and actions are part of the fabric of history of the market, the savannah, and the small pond.

Natural rationality is bounded by the very nonstationarity of the worlds we cocreate as we coexapt.

Technology Graphs and Economic Webs

Life takes its unexpected turns. I have been an academic scientist, a biologist, for thirty years at the University of Chicago, the National Institutes of Health, the University of Pennsylvania, then twelve stunningly exciting years at the Santa Fe Institute. After thirty years, I've written the canonical hundred or more scientific articles, was fortunate enough to receive a MacArthur Fellowship, during whose five years my IQ went up and then slumped back to normal as the funding ended, invented and patented this and that, and published two previous books of which I am proud, *Origins of Order* and *At Home in the Universe*, both by Oxford University Press.

I thought *Origins* and *At Home* were largely about the challenge of extending Darwinism to an account of evolution that embraced both self-organization and natural selection in some new, still poorly understood marriage. One hundred and forty years after Darwin, after all, we still have only inklings about the kinds of systems that are capable of adaptation. What principles, if any, govern the coevolutionary assembly of complex systems such as ecosystems or British common law, where a new finding by a judge alters precedent in ways that ricochet in small and large avalanches through the law? If new determinations by judges did not have any wider impact, the law could not evolve. If every new determination altered interpretation of precedents throughout the entire corpus of common law, the law also could not evolve.

My rough bet is that systems capable of coevolutionary construction, such as British common law, can evolve and accumulate complexity because they are somehow self-organized critical, and a power law distribution of avalanches of implications of new precedent ricochet in the law and in other complex coevolving systems to allow complexity to accumulate. Indeed, based on self-organized criticality, and more particularly on the analysis of the *NK* fitness landscape model discussed in *Origins* and *At Home* and the "patches" version of the *NK* model discussed in *At Home*, I am rather persuaded that adapting systems can best exploit the trade-off between exploitation and exploration at a rough phase transition between order and chaos. Here power law distributions of small and large avalanches of change can and do propagate through the system as it adapts.

So saying, and having published *Origins* and then *At Home*, I was rather surprised to find business people approaching me. The consulting companies of

McKinsey, Coopers and Lybrand, Anderson, and Ernst and Young began visiting the Santa Fe Institute to learn about the "new sciences of complexity." In due course, Chris Meyer at the Center for Business Innovation at Ernst and Young asked me if I might be interested in forming a fifty-fifty partnership with E and Y to discover if complexity science, that nuanced term, could be applied in the practical world. I found myself deeply intrigued. Was the work of my colleagues and myself mere theory or did it have application in real biospheres and econospheres? Why not plunge in and try my best to find out, to do it right, even knowing how early was the stage of the science we had been inventing.

Bios Group Inc., the partnership with Ernst and Young, is now just three-and-a-half years old. We have grown to over seventy people, heading, we hope, for a hundred. Our annual revenues are running at $6 million. We have a hopeful eye on $7 to $8 million this year with clients ranging from Texas Instruments, for whom we invented a novel adaptive chip, to the U.S. Marine Corps with its concern for adaptive combat, to Unilever, the NASDAQ stock market, Honda, Boeing, Johnson and Johnson, Procter & Gamble, Kellogg, Southwest Airlines, the Joint Chiefs of Staff, and others. We have spun out a biotechnology company, CIStem Molecular, that aims to clone the small cis acting DNA regions that control turning genes on and off in development and disease; a European daughter company, Euro-Bios; as well as EXA, a company spun out with NASDAQ to make tools for financial markets. I'm deeply glad to be chairman of the board and chief scientist of Bios, to be working with a very creative group of colleagues, and to be finding routes in the practical world where our ideas do, in fact, apply.

I mention Bios and my involvement because some of the science we have done bears on diverse aspects of practical economics and even begins to suggest pathways beyond the limitations of the Arrow-Debreu theory. I begin with Boeing, which came to Bios wondering how to design and build airplanes in a year rather than seven years. Out of what modular parts and processes, wonder Boeing folks, might it be possible to assemble a family of related aircraft for a diversity of markets?

The obvious approach was to invent "Lego World." As founding general partner and chief scientist, I duly authorized the expenditure of $23.94 to buy a largish box of Lego parts. (In truth, I fib. I actually won the Lego box at our first Bios Christmas party.)

Most of the readers of this book will be familiar with Lego. It is a construction game consisting of snap-together, plastic parts, based on square blocks that graduate in size, for example, 1 x 1, 1 x 2, 2 x 3, and 3 x 4 blocks. The blocks can be assembled into wonderfully complex structures, as many delighted children and adults have discovered.

But what might Lego World be? What, indeed. Well, consider a large pile of Lego blocks on a bare wooden table. Consider these blocks the "primitive parts." Now consider all the primitive construction or deconstruction operations, pressing two parts together or adding a primitive part to a growing assemblage, or taking a part off another part or off an assemblage.

Consider the pile of unassembled bare Lego parts as the founder set, and place in "rank 1" all the unique Lego objects that can be constructed from the founder set in a single construction step. Thus, a 1 x 3 can be attached in a specific overlapping way to a 2 x 4. Now place in rank 2 all the unique Lego objects that can be constructed from the founder set in two construction (or deconstruction) steps. Similarly, consider ranks 3, 4, 5, . . . , 20, 100, 10,000, 11,343,998, ...

A set of primitive parts and the transformations of those parts into other objects is a "technology graph." In fact, a technology graph is deeply similar to a chemical-reaction bipartite graph from a founder set of organic molecules, where the molecules are the objects and the reaction hyperedges linking substrate and products are the transformations among the objects. The graph is "bipartite" because there are two types of entities, nodes, and hyperedges, representing objects and transformations.

The first thing to notice about the Lego World technology graph is that it might extend off to infinity, given an infinite number of primitive Lego parts.

The second thing to notice is that within Lego World an adjacent possible relative to any actual set of primitive and more complex Lego structures is perfectly definable. The adjacent possible is just that set of unique novel objects, not yet constructed, that can be constructed from the current set of Lego objects in a single construction step. Of course, within the limited world of Lego we can think of the technologically adjacent possible from any actual. A Lego economy might flow persistently from simple primitive objects into the adjacent possible, building up evermore complex objects.

A third feature is that we might consider specific Lego machines, made of Lego parts, each able to carry out one or more of the primitive gluing or ungluing operations. Lego World could build up the machine tools to build other objects including other tools.

Indeed, in *Origins of Order* and *At Home in the Universe,* I borrowed theoretical chemist Walter Fontana's algorithmic chemistry and defined a mathematical analogue of Lego World, namely a "grammar model" of an economy. In that model, binary symbol strings represented goods and services, as do Lego objects in Lego World. In a grammar model, "grammar" specifies how symbol strings act on symbol strings, rather like machines on inputs, to produce new symbol strings. In Lego World, the grammar is specified by the ways primitive blocks can be attached or unattached and by any designation of which Lego objects can carry out which primitive construction operations. The grammar in question may be simple and "context-insensitive" or a far richer "context-sensitive" grammar in which what objects can be added in what ways to different objects depends upon the small or large context surrounding those blocks. In short, in a context-sensitive grammar, the objects and transformations rules are sensitive to the context of the objects and previous transformations themselves.

Before proceeding with current uses of Lego World and its intellectual children, notice that Lego World, like the grammar models in *Origins* and *At Home*, can become the locus of an economy, in which the sets of goods and services can expand over time and in which speciation and extinction events occur. In *Origins* and *At Home*, I built upon a suggestion of economist Paul Romer, and specified that each symbol string—or here, each Lego object—has some utility to a single consumer. The utility of each object to the consumer is subjected to exponential discounting over time. A Lego house today is worth more than a Lego house tomorrow and still more than a Lego house two days from now. And for simplicity's sake, the total utility of a bundle of goods is the sum of their discounted utilities to the consumer.

Next, I invoked a "social planner." The task of the social planner is to plan a pattern of production activities over time that optimizes the discounted happiness of the consumer. A standard approach is to adopt a finite planning horizon. The social planner thinks ahead, say, ten periods, finds that pattern of construction activities over time that creates the sequence of symbol-string goods, or Lego objects, that maximizes the time-discounted happiness of the consumer. Then, the social planner initiates the first-period plan, making the first set of objects. Next, the planner considers a ten-period planning horizon from period 1 to period 11, deduces the optimal second-period plan, taking account of the newly considered eleventh period, and carries out the second-period plan.

Because the total utility to the consumer is a simple sum of the discounted utilities of all the possible goods in the economy, finding the optimal plan at any period is just a linear programming problem, and the results are a fixed ratio of construction activities of all the objects produced at that period. The fixed ratio of the activities is the mirror of price, relative to one good, taken as the arbitrary "currency," or "numeraire."

Over time, the model economy ticks forward. At each period, in general, only some of the possible goods and services are constructed. The others do not make the consumer happy enough. Over time, new goods and services come into existence, and old ones go out of existence in small and large avalanches of speciation and extinction events.

Thus, a grammar model, or a physical instantiation of a grammar model such as Lego World, is a toy world with a technology graph of objects and transformations. With the addition of utilities to the different objects for one consumer or a set of consumers and a social planner—or more generally, with a set of utilities for the objects that may differ among consumers and with different costs and scaling of costs with sizes of production runs of different Lego objects—a market economy can be constructed. With defined start-up costs, costs of borrowing money, and bankruptcy rules, a model economy with an evolving set of goods and services can be created and studied.

In general, such economies will advance persistently into the adjacent possible. And because the number of unique Lego objects in each rank is larger than the number in the preceding rank, the diversity of opportunities and objects tends to increase as ever more complex objects are constructed.

More generally, we need to consider "complements" and "substitutes." Screw and screwdriver are complements; screw and nail are substitutes. Complements must be used together to create value; substitutes replace one another. Rather obviously, the complements and substitutes of any good or service constitute the economic niche in which that good or service lives. New goods enter the economy, typically, as complements and substitutes for existing goods. There is just no point in inventing the channel changer before the television set is invented and television programming is developed.

An economic web is just the set of goods and services in an economy, linked by red lines between substitutes and green lines between complements.

As we have seen, over the past million years, and even the past hundred years, the diversity of the economic web has increased. Why? Because, as in Lego World, the more objects there are in the economy, the more complement and substitute relations exist among those objects, as well as potential new objects in the adjacent possible. If there are N objects, the number of potential complement or substitute relations scales at least as N squared since each object might be a complement or substitute of any object. Thus, as the diversity of the objects in the web increases, the diversity of prospective niches for new goods and services increases even more rapidly! The very diversity of the economic web is autocatalytic.

If this view is correct, then diversity of goods and services is a major driver of economic growth. Indeed, I believe that the role of diversity of goods and services is the major unrecognized factor driving economic growth. Jane Jacobs had made the same point in her thoughtful books about the relation between economic growth and economic diversity of cities and their hinterlands. Economist Jose Scheinkman, now chairman of economics at the University of Chicago, and his colleagues studied a number of cities, normalized for total capitalization, and found that economic growth correlated with economic diversity in the city. In a similar spirit, microfinancing of a linked diversity of cottage businesses in the third world and the first world seems to be achieving local economic growth where more massive efforts at education and infrastructure, Aswan dams and power grids, seem to fail.

Indeed, in the same way in an ecosystem, organisms create niches for other organisms. I suspect, therefore, that over the past 4.8 billion years, the growth of diversity of species is autocatalytic, for the number of possible niches increases more

rapidly than the number of species filling niches. And in the linking of spontaneous and nonspontaneous processes, the universe as a whole advances autocat-
alytically into its adjacent possible, driven by the very increase of diversity by which
novel displacements from equilibrium come into existence, are detected, are coupled to, and come to drive the endergonic creation of novel kinds of molecules and
other entities. Economic growth is part and parcel of the creativity of the universe
as a whole.

Think of the Wright brothers' airplane. It was a recombination between an air-
foil, a light gasoline engine, bicycle wheels, and a propeller. The more objects an
economy has, the more novel objects can be constructed. When we were working
with rough stone in the Lower Paleolithic, we pretty much mastered everything
that could be done until pressure flaking came along. Most forms of simple stone
tools that could be made were made. Today, the adjacent possible of goods and
services is so vast that the economy, stumbling and lunging into the future adjacent
possible, will only construct an ever smaller subset of the technologically possible.

The economy is ever more historically contingent... As the biosphere is ever
more historically contingent... As, I suspect, the universe is ever more historically
contingent.

We are on a trajectory, given a classical $6N$-dimensional phase space, where the
dimensionality of the adjacent possible does seem to increase secularly and the
universe is not about to repeat itself in its nonergodic flow.

A fourth law?

I now discuss an algorithmic model of the real economic web, the one outside
in the bustling world of the shopping mall, of mergers and acquisitions. While
powerful, however, no algorithmic model is complete, for neither the biosphere
nor the econosphere is finitely prestatable. Indeed, the effort to design and construct an algorithmic model of the real economic web will simultaneously help us
see the weakness of any finite description.

It all hangs on object-oriented programming.

A case in point is the recent advent of Java, an object-based language, which, as
of February 1998 had a library of some eighty thousand Java objects. Goodness
knows how fast this library of objects is growing. Among the Java objects are "carburetor" objects, "engine block" objects, and "piston" objects, and objects come
with "functional" descriptors, such as "is a," "has a," "does a," "needs a," "uses a."

Both implicitly and, with modest work, explicitly, the "piston" object can discover that it fits into the cylinder hole in the "engine block" object to create a completed piston in a cylinder. The "carburetor" object can discover that it is to be
located on top of the "engine block" object, connected to certain "gas line" objects
in certain ways.

The physical engine block and piston, in reality, are complements, used together to create value. Thus, the representations of the engine block and piston as

algorithmic Java objects, together with algorithmic "search engines" to match the corresponding "is a," "has a," "does a," functions of complements and even substitutes, can as a matter of principle—and practicality—create an image of the real complements and substitutes in the real economic web.

In a fundamental sense, an appropriate set of Java objects, together with search engines for complements and substitutes matching "is a," "has a," "does a," constitutes a grammar of objects and linkings or transformations among objects. The grammar may be context independent or context sensitive or richer.

In short, properly carried out, Java objects and the proper search engines can create a technology graph of all the objects and transformed objects constructible from any founder set of objects. The Java objects are like Lego World. And Lego World, stated in terms of building simple combinatorial objects, is logically similar to a set of objectives that must be achieved by a military force to carry out its total objective. Entities and operations are deeply similar, as we will explore further below. Technology graphs concern objects and actions, things and objectives, products and processes in a single framework.

Therefore, in principle, we have begun to specify a means to characterize large patches of the global economic web. Let each of very many firms, at different levels of disaggregation, create Java objects proper to their activities, building materials, partially processed materials, partially achieved objectives, and objectives including products. Let these Java objects be characterized by appropriate "is a," "has a," "does a," lists, with adequate search engines looking for complements and substitutes. The result is a distributed web of Java objects linked functionally in at least many or most of the ways that are actually in use as complements and substitutes creating the millions of different goods and services in the current economy.

Much is of interest about such data on the real economic web. Among other features, any such graph has graph-typical characteristics. Some goods are central to the web, the car, computer, and so forth. Others are peripheral, such as the hula hoop and pet rock. Presumably, location of its products in the web structure has a great deal to do with the strategic position of a firm.

But there is more, for the economic web states its own adjacent possible. Given the Queen Mary and an umbrella, the umbrella placed in the smoke stack of the Queen Mary is in the adjacent possible. Not much use. But what about a small umbrella on the back of a Cessna 172 that opens upon landing: Ah, an air brake is a possible new good in the adjacent possible a few steps from here.

And still more. What are the statistics of the transformation of an economic web over time as new goods and services arise in the niches afforded by existing goods and services, and drive old goods and services extinct? No one makes Roman siege engines these days. Cruise missiles do the job better.

I believe such object-based economic web tools will come into existence in the near future. Indeed, Bios Group is involved in inventing and making them. And I believe that such tools will be very powerful means of coordinating activities within supply chains and within the larger economy when linked by automated markets.

But I do not believe any such algorithmic tool can be complete. Consider the case of the engineers discovering that the engine block is so rigid that the block itself can serve as the chassis for the tractor they are trying to invent. That exaptation seems a genuine discovery. Now imagine that we had had Java object models of all the parts that were to go into the tractor: engine block objects, carburetor objects, piston objects. If, among the properties of the engine block—the proud "is a," "has a," "does a," features—we had not listed ahead of time the very rigidity of the engine block or if that rigidity was not deducible from the other listed properties, then my vaunted economic web model with its algorithmically accessible adjacent possible could not have ever come up with the suggestion: Use the engine block, due to its rigidity, as the chassis.

You see again that unless there is a finite predescription of all the potentially relevant properties of a real physical object, our algorithmic approach can be powerful, but incomplete. Yet I cannot see how to construct such a finite predescription of all the potentially relevant properties of a real physical object in the real universe.

The world is richer than all our dreams, Horatio.

I must say to Arrow and Debreu, "Gentlemen, the set of goods and services is not finitely prestatable, so fixed-point theorems are of limited use."

And to my economist colleagues: Consider the economy as forever becoming, burgeoning with new ways of making a living, new ways of creating value and advantages of trade, while old ways go extinct. This too is the proper subject for your study, not just allocation of scarce resources and achievement of market-clearing prices. The economy, like the biosphere, is about persistent creativity in ways of making a living.

I find it intriguing to note certain parallels from our prior discussion of autonomous agents and propagating organization. At the level of molecular autonomous agents, I made the point repeatedly that work is the constrained release of energy and that autonomous agents do carry out work to construct the constraints on the release of energy such that the energy is released along specific channels and such that specific couplings of nonequilibrium energy sources to propagating organization arise. Think then of the role of laws and contracts, whose constraints enable the linked flow of economic activities down particular corridors of activities. The web of economic activities flows down channels whose constraints are largely legal in nature. The coming into existence of the enabling constraints of law is as central to economic development and growth as any other aspect of the bubbling activity.

Robust Constructibility

My first purpose in investing in an entire box of Legos was to explore and define concepts of "robust constructibility." We have succeeded, but run into fascinating problems of a general phase transition in problem solvability. In turn, this very phase transition suggests that in a coconstructing biosphere or econosphere rather specific restrictions arise and are respected by critters and firms, creatures and cognoscenti.

Recall the Lego founder set, and the rings, rank 1, rank 2, . . . , rank 11,983, . . . each containing the unique Lego objects first constructible from the founder set in a number of steps equal to the rank of that ring. Suppose a given Lego house is first constructible in twenty steps, hence, lies in rank 20. Now, it might be the case that there is a single construction pathway from the founder set to the Lego house in twenty steps. It might also be the case that there are thousands of construction pathways to the Lego house in twenty steps. In the latter case, intuitively, construction of the Lego house is robust. If one way is blocked, say because 1 x 3 blocks are temporarily used up, then a neighboring pathway will allow the Lego house to be constructed without delay, that is, in twenty steps, using other block sizes.

A related sense of robustly constructibility concerns how the number of ways to construct the Lego house increases if we take more than the minimum twenty steps, say, twenty-one, twenty-two, twenty-three, . . . steps. The number of ways may not increase at all or very slowly or hyperexponentially. If the number of ways increases very rapidly, it might be worth using twenty-two steps to make the Lego house, for it would be virtually impossible to block construction even if several types of building blocks and machines were temporarily broken.

But recall Boeing's question. They wanted to build a family of related objects. Hence, let us define still another related sense of robustly constructible.

Consider a family of Lego objects, a house, and a house with a chimney. Now consider each of the many ways from the founder set to build the Lego house. For each such way to build the Lego house, consider how to change construction minimally in order to build the Lego house with the chimney. Perhaps the chimney can just be added to the completed Lego house. More likely, it would be necessary to partially deconstruct that completed Lego house, then go on to construct the house with the chimney. So there is a last branch point during construction on the way both to the Lego house and the Lego house with the chimney.

The branch point object and/or operation that is simultaneously on the way to the house and the house with the chimney is an interesting intermediate-complexity object or operation because it is polyfunctional. It can be used in at least two further ways.

When we build a house, we all know that boards and nails are primitive objects and the completed house is the finished object. But some intermediate objects, say, framed windows and framed walls, are commonly used. Why? Because they are intermediate objects that are polyfunctional. The technology graph and its branch points are identifying the intermediate-complexity polyfunctional objects for us.

But things are more subtle. It might be the case that from the last branch point on a way to make both the house and the house with the chimney there is only a single pathway forward to the house and there is only a single pathway forward to the house with the chimney. Not robust. Stupid stopping spot. Either pathway can readily be blocked. Suppose instead we consider an intermediate object three steps prior to the last branch point on the way outward from the founder set. Ah, perhaps there are thousands of ways to complete the house and to complete the house with the chimney. Any single blockage or small set of blockages is readily overcome. The house, or the house with the chimney, can be built without delay if all 1 x 2 blocks are temporarily out of stock. Now this is a smart, robust, intermediate-complexity polyfunctional object-objective. And it may cost no more to stockpile such smart intermediate objects!

So, here is a new view of process design and inventory control.

Bios colleague Jim Herriot has made a delightful little Java computer model to show technology graphs in action. The program shows a "chair" object, a "seat" object, a "back" object, and a "leg" object. In addition, there are "foam" and "padding" objects, two "attachment" objects, a set of "screw" objects, "nail" objects, "wood" objects, a "saw" object, a "hammer" object, and a "screwdriver" object. Each object comes with its own characteristic set of "is a," "has a," "does a," features.

The program assembles coherent technology graphs and chair-assembly pathways as follows: An object tries a connection, shown by a black line, to another object. In effect, the chair object extends a line to the screw object as it says, "I need a lean-on! I need a lean-on!" The screw object responds, "I do twist holds, I do twist holds!" There is no match. After many random tries, the "chair" object extends a black line to the "back" object. "I need a lean-on, I need a lean-on," says the "chair" object. "I do lean-ons, I do lean-ons," cries the "back" object with cybernetic joy. The black line becomes a yellow line as a contract is signed between the "chair" and "back" objects. In a similar way, the "back" object needs "padding" and "wood" objects, and either the complementary pair "nail and hammer" objects or their substitutes, "screw and screwdriver" objects, to carry out an "attachment" operation.

In due course, the conditions are met to begin construction of partial objects on the way to the entire chair. Legs, then backs, begin to be assembled as screws and nails are used up. Eventually, a seat is constructed too, and the first chair triumphantly follows.

All goes well until all the screws are used up. The seat, having relied on screws and screwdrivers to attach padding, goes nuts and looks about frantically, asking the screwdriver to work on nails. No luck. Eventually, the seat tries nails and hammers jointly, and that complementary pair works. More chairs are constructed, then nails run out and failures propagate throughout the system.

Not a metaphor, the technology graph. Rather, a new tool to understand the general principles of robust constructibility, the structure of economic webs, a knowledge-management tool for a firm, a new hunk of basic science. Indeed, one of the interesting features of technology graphs is that they constitute the proper conceptual framework to consider process and product design simultaneously. As far as I can tell, we have not had such a conceptual framework before.

Nor is the technology graph limited to manufacturing. The same general principles apply, for example, in military or other logistic operations. Technology graphs, in these other contexts, become the sequential set of requirements needed to meet subobjectives that robustly culminate in the achievement of an overall objective. Taking hill 19 after diverting gracefully from orders to take hill 20 is logically related to making the house with the chimney after starting to make the house without the chimney.

A Phase Transition in Problem Solvability

Part of the basic science of technology graphs stems from generic phase transitions in problem solvability in many combinatorial optimization or satisficing problems in biology and economics. I turn now to discuss these generic phase transitions.

I begin with a metaphor. You are in the Alps. A yellow bromine fog is present. Anyone in the fog for more than a microsecond will die. There are three regimes: the "dead," the "living dead," and the "survivable."

The "dead": The bromine fog is higher than Mont Blanc. Unfortunately, everyone dies.

The "living dead": The bromine fog has drifted lower and Mont Blanc, the Eiger, and the Matterhorn jut into the sunlight. Hikers near these three peaks are alive.

But consider that even mountains are not fixed. Plate tectonics can deform the mountainous landscape. Or, in the terms of the last chapter, the mountainous fitness landscape of a species or a firm or an armed force can change and persistently deform due to coevolution when other species or firms or adversaries change strategies. If the mountainous landscape deforms, Mont Blanc, the Eiger, and the Matterhorn will eventually dip into the bromine fog. As this happens, perhaps new peaks jut into the sunshine. But those peaks will typically be far away from Mont Blanc, the Eiger, and the Matterhorn.

Alas, the hikers near those three initial peaks will die as they are dipped into the lethal fog and are too far from the newly emerged sun drenched peaks to reach them. This "isolated peaks regime" is the living dead regime.

But there is a third regime a phase transition away.

The survivable regime: Let the bromine fog drift lower. More and more peaks jut into the sunshine. At some point, some magical point, as more peaks emerge into the sunshine, quite suddenly, a hiker can walk all the way across the Alps in the sunshine.

This is a phase transition from the isolated peaks regime. A connected web—mathematically, a percolating web—of connected "solutions" has emerged suddenly as the fog lowers. Now consider hikers striding across the Alps, knapsacks and hearts full. If plate tectonics rather slowly deforms the landscape, then whenever hikers are about to be dipped into the lethal bromine fog, they can take a sideways step in some direction and remain in the sunshine.

The percolating web of solutions regime is persistently survivable. In fitness landscape terms, if you are evolving on a fitness landscape that is deforming and are in your survivable regime, you can continue to exist by stepping somewhere from wherever you happen to find yourself.

This phase transition is generic to hard combinatorial optimization problems. A case in point is the well-known job shop problem. The job shop problem posits M machines and O objects. The idea is to build the O objects on the machines. Each object requires being on each machine in some set order for some fixed period of time. Perhaps object 1 must be on machine 13 for 20 minutes, then machine 3 for 11 minutes, then machine 4 for 31 minutes. In turn, object 2 must be on machine 1 for 11 minutes, then machine 22 for 10 minutes, and so on.

A schedule is an assignment of objects to machines such that all objects are constructed. The total length of time it takes to construct the set of objects is called the "makespan." We may consider, for each schedule, a definition of neighboring schedules, such as swapping the order in which two objects are assigned to a given machine. Given the set of possible schedules, the neighborhood relation between schedules, and the makespan of each schedule, there is a makespan fitness landscape over the space of schedules of the job shop problem.

We want to minimize makespan. But to keep our mountainous landscape metaphor where high peaks are good, let us consider optimizing efficiency by minimizing makespan. So schedules with low makespan correspond to points of high efficiency on the job-shop fitness landscape. Clearly, short makespan makes the problem hard, long makespan makes the problem easy. So long makespan is like the bromine fog being low, while short makespan is like the bromine fog being high.

Does the phase transition occur in typical job shop problems as makespan is tuned from long to short? Figure 9.2 shows the results Vince Darley obtained. The figure plots makespan, short to long, on the x-axis and the number of schedules at a given makespan on the y-axis.

As you can see, for long enough makespan, as makespan decreases there are roughly a constant number of schedules at each makespan. But at a critically short makespan, the number of solutions starts to fall abruptly. The corner where the curve starts to turn is the phase transition between the survivable regime for longer makespans and the isolated peaks/living dead regime for shorter makespans. (I cheat slightly, the sharpness of the corner increases as the size of the job shop problem increases in numbers of machines, M, and objects, O.)

There are a number of direct tests for this phase transition. In the survivable regime, at longer makespans and lower efficiency than the phase transition makespan, start with a given schedule at a given makespan. Now examine all "nearby" schedules and test if any is of an equal or better makespan. Continue to "walk" across the space of schedules via neighbors to test if there is a connected web of schedules with makespans at least as good as our initial schedule's makespan. Either the web percolates across the space of solutions or it does not. If the web percolates, then the initial schedule was in the survivable regime. If only isolated regions of neighboring schedules of the same approximate makespan are found, then you are in the isolated peaks regime.

A second test looks at the "Hausdorf dimensionality" of the acceptable solutions at a given makespan or better. The Hausdorf dimension is computed by considering an initial schedule at a given makespan, then by considering from among all the 1-mutant neighbor schedules the number of them that are of the same or better makespan as the initial schedule's makespan, then doing the same among all the 2-mutant neighbor schedules. The Hausdorf dimension of the acceptable set of schedules at that makespan and point in the job shop space is the ratio of the loga-

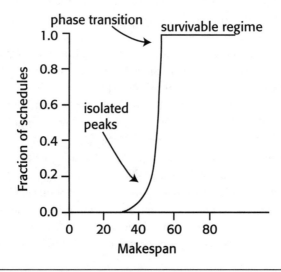

FIGURE 9.2 The transition from the isolated peaks regime to the survivable regime in a job shop problem as makespan increases. Robust survivable operations occur on the horizontal region of the curve near the phase transition region where the curve bends sharply downward as makespan decreases. Phase transition becomes sharper as size of job shop problem increases.

rithm of the 2-mutant acceptable schedules to the logarithm of the 1-mutant acceptable schedules. In effect, the Hausdorf dimension shows how rapidly—in how many dimensions of the job-shop schedule space—acceptable schedules of a given makespan or better are growing. In the survivable regime, the Hausdorf dimension, on average, is greater than 1.0. In the isolated peaks regime, averaged over the job shop space, the Hausdorf dimension is less than 1.0. At the phase transition, the dimensionality is 1.0.

The phase transition I have just noted is generic for many or most hard combinatorial optimization problems. It is not true for all fitness landscapes. For example, it would not hold on a conical Fujiyama landscape. But the Fuji landscape corresponds to a simple, typically linear, optimization problem. Hard combinatorial optimization problems are multipeaked due to conflicting constraints.

A further interesting connection relates the statistics of search on the job shop landscape to learning curves in economics. Learning curves, well known in arenas from airplane manufacture to diamond cutting to cigar manufacture, show that every time the total output of a plant is doubled, the cost per unit falls by a rough constant percentage, typically 5 to 10 percent. If the logarithm of the cost per unit is plotted on the y-axis and the logarithm of the cumulative number of units produced is plotted on the x-axis, one gets a typical straight-line power law that decreases downward to the right.

The fascinating thing is that this feature probably reflects the statistics of search for fitter variants on rugged, correlated fitness landscapes such as the job shop problem. Typically, in such problems, every time a fitter 1-mutant variant is found, the fraction of 1-mutant variants that are still fitter falls by a constant fraction, while the improvement achieved at each step is typically a constant fraction of the improvement achieved at the last step. These properties yield the learning curve. My colleagues Jose Lobo, Phil Auerswald, Karl Shell, Bill Macready, and I have published a number of papers on the application of the statistics of rugged landscapes and learning curves.

But there is another even more important point. We can control the statistical structure of the problem spaces we face such that the problem space is more readily solvable. We can, and do, tune the structure of the problems we solve. The capacity to tune landscape structure shows up in the job shop problem. In most cases, improving the structure of the problem space requires relaxing conflicting constraints to move the problem into the survivable regime at the makespan, or efficiency, you require. For a specific case, in my statement of the job shop problem, I asserted that the O objects must each have access to the M machines in some fixed order.

Now simple observation and experience tells you that you can put on your shirt and pants in either order, but you had better put on your socks before your shoes. In other words, some steps in life are permutable, others are not. Suppose in our job shop problem, a fixed fraction, P, of the steps in the construction of each of the O objects were jointly permutable. As P increases and more steps are permutable, the conflicting constraints in the total problem are reduced. But this in turn means that the entire space of solutions improves, that is, the entire space shifts toward lower makespan, or higher efficiency.

And now the relation to the bromine fog metaphor can be stated clearly. As the number of permutable steps increases, the conflicting constraints are reduced. The entire makespan-efficiency landscape is lifted higher, the peaks are higher, and the landscape, with fewer conflicting constraints, is smoother. All in all, the result is that the percolating web of solutions where the hikers can walk all across the Alps occurs at a higher efficiency and shorter makespan.

In terms of Figure 9.2, the capacity to permute steps in the job shop problem shifts the phase transition point leftward, toward shorter makespan. Equivalently, if one wants to shift the curve in Figure 9.2 to the left, purchase a computable number of machines that are polyfunctional so that the same machine can be used for more than one job. That too reduces conflicting constraints.

There is another view of this generic phase transition between solvable and nonsolvable that again highlights the role of having alternative ways of doing things, robustly constructible strategies that are not easily blocked. In addition, the same simple model, the Ksat model discussed in the previous chapter, begins to account for at least the following anecdotal observation, which is apparently typical: Colleagues at Unilever noted to us that if they have a plant that manufactures different types of a product, say, toothpaste, then the plant does well when the diversity of products grows from three to four to ten to twenty to twenty-five, but at twenty-seven different toothpastes, the plant suddenly fails. So rather abruptly, as product diversity in a given plant increases, the system fails.

Why? Presumably it is the same phase transition in problem solvability.

Consider again the Ksat problem, taken as a model for community assembly in the last chapter based on the work of Bruce Sawhill and Tim Keitt. Figure 8.13 repeats the Ksat problem and shows again the phase transition.

Recall that a Ksat problem consists in a logical statement with V variables, in C clauses, in normal disjunctive form: $(A_1 \lor A_2)$ and $(A_3 \lor A_4)$ and (not $A_2 \lor A_4$). As we discussed in the previous chapter, a normal disjunctive expression with C clauses, V variables in total, and K variables per clause is satisfiable if there is an assignment of true or false to each of the V variables, such that the expression as a whole is true.

The normal disjunctive form makes it clear that as there is an increase in the number of alternative ways, K, of carrying out a task, or making a clause true, it is easier to satisfy the combined expression. More generally, as noted in the last chapter and shown in Figure 8.13, there is a phase transition in the probability that a random Ksat expression with V variables, K per clause, and C clauses can be satisfied by some assignment of true or false to the V variables. The phase transition occurs in the horizontal axis, labeled C/V, which is the mean number of clauses in which any variable occurs. Obviously, as C/V increases, conflicting constraints increase. The phase transition from easily solvable to virtually impossible to solve occurs at a point on the C/V axis equal to log 2 x 2 raised to the K power, or 0.6 x 2^K. Hence, as K increases, the phase transition shifts outward to greater C/V values.

But Figure 8.13 gives us an intuitive understanding of Unilever's problem, in fact, the problem is far more general than Unilever's assembly plants. Think of each clause, $(A_1 \vee A_2)$, et cetera, as a way to make one of the toothpaste products, and think of the conjunction, $(A_1 \vee A_2)$ and $(A_3 \vee A_4)$ and..., as the way to make all twenty-seven toothpastes. Then as the number of clauses, hence toothpaste products, increases for a given plant with V variables to use in making these different products, all of a sudden the conjoint problem will have so many conflicting constraints that it will cross the phase transition from solvable to insolvable.

Moreover, for any given number of clauses and V variables, if a given assignment of true or false to the V variables satisfies the Ksat problem, we can ask if any of the 1-mutant neighbor assignments of true and false to the V variables that change the truth value assigned to one of the V variables also satisfy the Ksat problem. Thus, we can study the phase transition from a survivable percolating web of solutions regime when V/C is lower to an isolated peaks regime as V/C increases to virtual impossibility for high V/C.

Thus, just as in the job shop problem, as product diversity increases for a fixed plant, a phase transition from survivable to unsurvivable will occur because the conflicting constraints will increase with C/V. The resulting landscape becomes more rugged, the peaks lower, the yellow bromine fog rises from the survivable to the living dead to the dead regime, covering Mont Blanc and all hikers in the Alps (Figure 9.3).

But this takes us back to the technology graph and robust constructibility. Recall from above that the Lego house and Lego house with a chimney might be robustly constructible from wisely chosen intermediate-complexity objects on the pathway to both houses, with thousands of ways to get there. At no extra cost, we might choose to stockpile that intermediate-complexity polyfunctional object rather than another choice.

But intermediate-complexity polyfunctional objects are just what allows multiple pathways, multiple permutations of construction steps to our two final objects. Hence, these same smart intermediate objects reduce the conflicting constraints in the fitness landscape over the construction space to make our desired set of objects, or our set of toothpastes. Lowering the conflicting constraints makes the efficiency peaks of the fitness landscapes higher, hence, allows survivable operations at a higher level of product diversity.

Thus, by use of the technology graph to design both products and processes, we

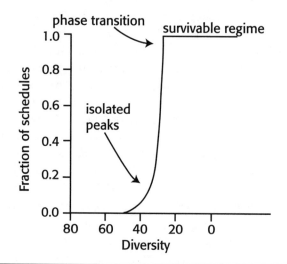

FIGURE 9.3 Transition from isolated peaks regime to survivable regime as the diversity of goods produced in a given facility decreases from high to low. Figure suggests a bound on diversity that can be manufactured in a given facility because conflicting constraints increase as diversity increases. Robust operations occur on the horizontal part of the curve just before the curve drops sharply as diversity of goods increases. The same concepts should apply to the complexity of a military campaign that can be waged in terms of the diversity of weapon systems—variables—and subobjectives—clauses. Robust operations should occur in the survivable regime.

can choose a family of products and construction pathways with highly redundant intermediate objects. That choice makes the problem space easy to solve rather than hard to solve. We have thereby tuned the statistical structure of our problem space into a survivable regime. Furthermore, we can test whether our choice of construction pathways to the house and/or house with a chimney is robustly survivable or in the living dead–isolated peaks regime. We need merely use the technology graph to test for percolating sets of 1-mutant neighboring pathways of construction of the same objects and the average Hausdorf dimension of such pathways.

No need to operate in the isolated peaks regime. Indeed, if you face loss of parts and machines, you had best locate back from the phase transition, deep enough into the survivable regime to survive. And if you are a military force fighting against an enemy whose strategy changes persistently deform your payoff landscape and whose efforts are to destroy your capacity to fight, you had best operate even further back from the phase transition in the survivable regime. Indeed, the normal disjunctive form in Figure 9.3 is a rough image of the complexity of a campaign you can fight—the number of clauses that must be jointly satisfied to meet your objectives, where each clause is a subobjective and there are K alternative ways to meet that objective using V weapon systems.

Just as warfare and the economy as a whole have much in common, warfare and the biosphere have much in common. If you are a species coevolving with other species, you had best operate back from the phase transition well into the survivable regime.

There is a message: If you must make a living, for God's sake, make your problem space survivable!

This brings us back to a point made in early chapters. Recall the no-free-lunch theorem proved by Bill Macready and David Wolpert. Given a family of all possible fitness landscapes, on average, no search algorithm outperforms any other search algorithm. Hill climbing is, on average, no better than random search in finding high peaks, when averaged over all possible fitness landscapes.

The no-free-lunch theorem led me to wonder about the following: We organisms use mutation, recombination, and selection in evolution, and we pay twofold fitness for sex and recombination to boot. But recombination is only a useful search procedure on smooth enough fitness landscapes where the high peaks snuggle rather near one another.

In turn, this led me to wonder where such nice fitness landscapes arise in evolution, for not all fitness landscapes are so blessedly smooth. Some are random. Some are anticorrelated.

In turn, this led me to think about and discuss natural games, or ways of making a living. Since ways of making a living evolve with the organisms making those livings, we got to the winning games are the games the winners play. Which led me to suggest that those ways of making a living that are well searched out and exploited by the search mechanisms organisms happen to use—mutation, recombination, and selection—will be ways of making a living that are well populated by organisms and similar species. Ways of making a living that cannot be well searched out by organisms and their mutation recombination search procedures will not be well populated.

So we came to the reasonable conclusion that a biosphere of autonomous agents is a self-consistently self-constructing whole, in which agents, ways of making a living, and ways of searching for how to make a living all work together to co-construct the biosphere. Happily, we are picking the problems we can manage to solve. Of course, if we could not solve our chosen ways to make livings, we would be dead.

And there is, I think, a molecular clue that the biosphere is persistently coconstructing itself in the survivable regime for a propagating set of lineages. We have just characterized the survivable percolating web regime where the fitness landscape of each creature deforms due to the adaptive moves of other creatures, but there are always neighboring ways of surviving. Genetically, those neighboring ways are one or a few mutations or recombinations away from where the species population is right now. If the biosphere has coconstructed itself such that most species are in a survivable regime, then as coevolution occurs, most species will persist but may well transform, for example, to daughter species. One would guess that this mildly turbulent process is rather continuous, perhaps with some self-organized critical bursts on a power law scale.

The "molecular clock hypothesis" seems to fit these facts. If one compares hemoglobins from humans, chimps, horses, whales, and so on, in general, the longer ago we diverged from one another in the evolutionary record, the more amino acid mutations distinguish the hemoglobins of the two species involved. Our hemoglobin is very similar to chimp hemoglobin and quite different from the whale. So good is this correlation that, within given protein families, it is argued that mutations accumulate with timelike clockwork, hence the molecular clock hypothesis, in which a number of amino acid differences can be taken as a surrogate for time from the most common ancestor. Different protein family clocks seem to run at different rates.

There is evidence that the clock does not run quite smoothly. John Gillespie, a population biologist now at the University of California at Davis, showed some years ago that amino acid substitutions seemed to come in short bursts that accumulate over long periods of time to a rough molecular clock that "stutters." Gillespie argued that fitness landscapes were episodically shifting and the bursts of amino acid substitutions were adaptive runs toward nearby newly formed peaks. I agree and suggest that the near accuracy of the molecular clock data over hundreds of millions of years and virtually all species strongly suggests that the biosphere has coconstructed itself such that species, even as they speciate and go extinct, are, as lineages, in the persistently survivable regime.

We as organisms have, in fact, constructed our ways of making a living such that those problem spaces are typically, but not always, solvable as coevolution proceeds. And, on average, the same thing holds for the econosphere. As old ways of making a living go extinct, new ones persistently enter. We too, it appears, have coconstructed our econosphere such that our ways of making a living, and discovering new ways of making a living, are manageable, probably in a self-organized critical manner, with small and large speciation and extinction events.

And there is a corollary: If you are lucky enough to be in the survivable regime, you can survive by being adaptable. What is required to be adaptable as an organism or organization? We discussed this in the last chapter. A good guess is that an organism or organization needs to be poised in an ordered regime, near the edge of chaos, where a power law distribution of small and large avalanches of change propagates through the system such that it optimizes the persistent balance between exploration and exploitation on ever-shifting, coevolving fitness landscapes.

Laws for any biosphere extend, presumably, to laws for any economy. Nor should that be surprising. The economy is based on advantages of trade. But those advantages accrue no more to humans exchanging apples and oranges than to root nodules and fungi exchanging sugar and fixed nitrogen such that both make enhanced livings. Thus, economics must partake of the vast creativity of the universe. Molecules, species, and economic systems are advancing into an adjacent possible. In all cases, one senses a secular trend for diversity to increase, hence for the dimensionality of the adjacent possible to increase in our nonergodic journey.

Perhaps again we glimpse a fourth law.

Chapter 10

A COCONSTRUCTING COSMOS?

F ROM BIOSPHERES to the cosmos? Yes, because they may share general themes. The major enquiry of *Investigations* has concerned autonomous agents and their coconstruction of biospheres and econospheres whose configuration spaces cannot be finitely prestated. These themes find echoes in thinking about the cosmos as a whole. But abundant caution: I am not a physicist, the problems are profound, and we risk nonsense.

Whatever the risk, two facts are true. First, since the big bang our universe has become enormously complex. Second, we do not have a theory for why the universe is complex. Equally unarguably, the biosphere has increased in molecular diversity over the past four billion years, just as the standing diversity of species has increased. And equally unarguably, the econosphere has become more complex over the past few million years of hominid evolution. We know this with confidence. If we lack a theory, it is not because the staggering facts of increasing diversity and complexity that stare us in the face do not deserve a theory to account for them.

But we have seen hints of such a theory in the coconstruction of biospheres and econospheres by the self-consistent search of autonomous agents for ways to make a living, the resulting exapting novel ways of making livings, the fact that new adjacent niches for yet further new species grow in diversity faster than the species whose generation creates those new adjacent possible niches, and the search mechanisms to master those modes of being. We have seen a glimmer of something like a fourth law, a tendency for self-constructing biospheres to enlarge their work-space, the dimensionality of their adjacent possible, perhaps as fast, on average, as is possible—glimmers only, not yet well-founded theory nor well-established fact. But glimmers often precede later science.

Consider again how the chemical diversity of the biosphere has become more diverse in the past four billion years, urged into its adjacent possible by the genuine chemical potential from the chemical actual into the adjacent possible, where the actual substrates exist and the adjacent possible products do not yet exist. Each time the molecular diversity of the biosphere expands, the set of adjacent possible reactions expands even faster. Recall our simple calculation that for modestly complex organic molecules, any pair of molecules could undergo at least one two substrate–two product reaction. But then the diversity of possible reactions is the square of the diversity of chemicals in the system. As the diversity of molecular species increases, there are always proportionally more novel reactions into the adjacent possible. If we take the formation of a chemical species that has never existed in the biosphere, or perhaps the universe, as a breaking symmetry, then the more such symmetries are broken, the more ways come into existence by which yet further symmetries may be broken.

And the chemical case makes clear the linking of the flows of matter and energy in this sprawling chemical diversity explosion. Many such reactions will link exergonic and endergonic processes. As this occurs, energy is pumped from the exergonic partner into the products requiring endergonic synthesis. These products—the chemical diversity in the bark of a redwood tree, for example—take their place in the chemical actual, poising the biosphere, and thus the universe, for its next plunge into the chemical adjacent possible.

Consider again equilibrium statistical mechanics. At its core, statistical mechanics relies on the same kind of statistical argument as does the flipping of a fair coin 10,000 times. We all understand that distributions of roughly 5,000 heads and 5,000 tails are far more probable macrostates than distributions with all heads or all tails. Now consider the general argument I have made that as molecular diversity increases, the diversity of reactions increases even faster, and that there is a genuine chemical potential from the actual into the adjacent possible. And consider again the general argument made just above that the greater the diversity of molecular species and reactions, the more likely the coupling of exergonic and endergonic reaction pairs driving the endergonic synthesis of new adjacent possible molecules that poise the system to advance again into the next adjacent possible. While the detailed statistical form of these chemical reaction graphs are not yet known, they too smell of "law." As in the case of fair coin flips and equilibrium statistical mechanics, it is as if here again the mathematical structure compels the consequent behavior of matter and energy. In the case of the nonergodic and nonequilibrium chemical flux into the adjacent possible, the universe is busy diversifying itself into myriad complexity.

The universe is enormously complex, and we don't really yet know why. May there be new ways of thinking of the cosmos itself? If a mere glimmer can be acceptable as potentially useful early science, then the burden of this chapter is to suggest perhaps, yes.

It is not obvious, in fact, that the universe should be complex. One can imagine universes governed by general relativity that burst briefly into big bang being, then recollapse in a rapid big crunch within parts of a second or a century. Alternatively, one can imagine universes governed by general relativity that burst into big bang being and expanded forever with no further complexity than hydrogen and helium and smaller particles in an open and ever-expanding dark, cold vastness.

Here we are, poised, it seems (but see below) between a universe that will expand forever and a universe that will eventually ease into gentle contraction, then rush to a big crunch.

Our fundamental theories in physics, and just one level up, biology, remain ununited. Einstein's austere general relativity, our theory of space, time, and geometry on the macroscale, floats untethered to quantum mechanics, our theory of the microscale, seventy-five years after quantum mechanics emerged in Schrödinger's equation form for wave mechanics. Theoretically apart, general relativity and quantum mechanics are both verified to eleven decimal places by appropriate tests. But it remains true that general relativity and quantum mechanics remain fitfully fit, fitfully un-united. And Darwin's view of persistent coevolution remains by and large unconnected with our fundamental physics, even though the evolution of the biosphere is manifestly a physical process in the universe. Physicists cannot escape this problem by saying, "Oh, that's biology."

The Complexity of the Universe

Why the universe is complex rather than simple is, in fact, beginning to emerge as a legitimate and deep question. In the past several years, I have had the pleasure to come to know Lee Smolin, whose career is devoted to quantum gravity and cosmology. Most of what I shall write in this chapter reflects my conversations and work with Lee and his colleagues, who have been wonderfully welcoming to this biologist. Sometimes outsiders can make serious contributions. Sometimes outsiders just make damned fools of themselves.

Caveat lector, but I will continue.

In Smolin's book, *The Life of the Cosmos,* he raises directly the question of why the universe is complex. Current particle physics has united three of the four fundamental forces—the electromagnetic, weak, and strong forces called the "standard model." With general relativity, which deals with the remaining force, gravity; this provides a consistent framework. Particle physics plus general relativity have altogether some twenty "constants" of nature, which are parameters of the standard model and general relativity, such as the value of Planck's constant, h; the fine structure constant, that is, the ratio of the electron rest mass to proton rest mass; the gravitational constant, g; and so forth. Smolin puts approximate maximum and minimum bounds on these twenty constants and asks a straightforward question: In a twenty-dimensional parameter space ranging over the plausible values of these twenty constants, what volume of that parameter space is consistent with values of the constants that would yield a complex universe with stars, chemistry, and potentially, life?

Smolin's rough answer is that the volume of parameter space for the constants of nature that would yield a complex universe are something like 10 raised to the minus 27th power. That is, a tiny fraction of the possible combinations of the values of the constants are consistent with the existence of chemistry and stars, as well as life. For the universe to be complex, the constants must be sharply tuned.

Smolin's argument could be off by many orders of magnitude without destroying his central point: The fact that our universe is complex, based on our current theories of the standard model and general relativity, is surprising, even astonishingly surprising.

Many physicists have remarked upon this fine tuning of the constants.

There have been several responses to this issue, some raised prior to Smolin's work. One is based on a view of multiple universes and the "weak anthropic principle." This principle states that there exist multiple universes, but only those universes that were complex would sport life forms with the wit to wonder why their universe was complex. So the very fact that we humans are here to wonder about this issue merely means that we happen to be in one of the complex universes among vastly many universes. The argument is at least coherent. But it's hard to be thrilled by this answer.

The "strong anthropic principle" goes further—indeed, too far—and posits that, for mysterious reasons, the universe is contrived such that life must arise to observe it and wonder at it. Few think the strong anthropic principle counts as science in any guise.

Smolin points out that there are two possible answers to the puzzle of the complexity of the universe. Either we will find a parameter-free description—a kind of supertheory—that yields something like our complex universe with its constants, or some historical process must pick those constants. Smolin proposes the possibility of "cosmic natural selection." Here, daughter universes are born of black holes. Universes with more black holes will have more daughter universes. Given minor heritable variation in the constants of the laws of the daughter universes, cosmic natural selection will select for universes whose constants support the formation of a near-maximum number of black holes. He then argues that on very crude calculations most alterations in the known constants would be expected to lower the number of black holes. Lee points out that his theory is testable, for example, by deducing that our constants correspond to near-maximum black hole production, and that his theory has not been ruled out yet.

I confess I am fond of and admire Lee Smolin a great deal, but I don't like his hypothesis. Why? Well, preferably, one would like a theory that had the consequence that any universe would be complex like ours and roughly poised between expansion and contraction. We have no such theory at present, of course. The remainder of this chapter discusses ideas and a research program that just might point in this direction.

As a start, we can begin with the most current view of the large-scale structure and dynamics of the universe. The most recent evidence suggests that on a large enough scale the universe is flat, the matter distribution is isotropic, and—a recent surprise—the universe may be expanding at an *accelerating* rate. This latest result, if it holds true, contravenes the accepted view of the past several decades that the rate of expansion of the universe has been gradually slowing since the big bang. The hypothesis that the universe is exactly poised between persistent expansion and eventual collapse has held that the rate of expansion of the universe will gradually slow, but never stop.

One way to explain a persistent accelerating expansion of a flat universe is to reintroduce Einstein's "cosmological constant" into the field equations of general relativity. A positive cosmological constant expresses itself as a repulsive force between masses that increases with the distance between those masses. Some physicists think that a positive cosmological constant must be associated with some new source of energy in free space. The source of such an energy is currently unknown.

Quantum Mechanics and Classicity

Before turning to the huge difficulties of quantum gravity, we should review the fundamental mystery of quantum mechanics. Most readers are familiar with the famous two-slit experiment, which exhibits the fundamental oddness of quantum interference. Feynman, in his famous three-volume lectures on physics, gives the mystery as simply as possible: We begin with a gun shooting bullets. The bullets pass through one of two holes in a metal plate and fly further on, landing on a flat layer of sand in a box. Bullets passing through either hole may be deflected slightly by hitting the walls of the hole. Thus, in the sandbox behind the metal plate, we would expect, and actually would find, two mounds of bullets. Each mound would be centered on the line of flight of the bullet from the gun through the corresponding hole to the sandbox, with a "Gaussian" or normal bell-shaped distribution of bullet densities falling away from the peak of each mound.

When we use monochromatic light rather than bullets, we note the following: If the light hits the sandbox, changed into a photon-counter surface, we find that the size of the energetic impact is the same each time a photon hits the surface. Photons of a given wavelength have a fixed energy. A photon either is recorded at a point on the surface or not. Whenever one is recorded, the full parcel of energy has been detected at the surface. Now if only one hole is open, one gets the Gaussian mound result. Most photons pass through the hole unscathed and arrive in a straight line at the photon-counter surface. A Gaussian distribution peaked at that center is present because some photons are deflected slightly by the edges of the hole.

But if two holes are open, then one gets the famous interference pattern of light and dark interfering circles spreading from the centers on the photon-counter surface that were the peaks of the mounds seen when hole 1 or hole 2 was open. Of course, as Feynman points out, there is no way to account for this oddness in classical physics.

Quantum mechanics was built to account for the phenomenon. The Schrödinger equation is a wave equation. The wave that propagates from the photon gun is an undulating spherically spreading wave of probability "amplitude." The amplitude at any point in space and time is the square root of the probability that the photon will be found located at that point. To obtain the actual probability, the amplitude must be squared.

A central feature of Schrödinger's equation is its linearity. If two waves are propagating, the sum and differences of those waves are also propagating. It is the essential linearity of quantum mechanics that makes the next puzzle, the link from quantum to classical worlds, so difficult. For a central puzzle of quantum mechanics becomes the relation between this odd quantum world of possible events, where the possibilities can propagate, but never become actual, and the classical world of actual events.

A variety of approaches to the liaison between the quantum and classical realms exist. The first is the "Copenhagen interpretation," which speaks of the "measurement event," when the quantum object interacts with a macroscopic classical object, the measuring device, and a single one of the propagating possibilities becomes actual in the measurement event, thereby "collapsing" the wave function. A second approach is the Everett multiworld hypothesis, which asserts that every time a quantum choice happens, the universe splits into two parallel universes. No one seems too happy with the Everett interpretation. And few seem very sure what the Copenhagen interpretation's collapse of the wave function might really mean.

Meanwhile, there are two other long-standing approaches to the link between the quantum and classical worlds. The first is Feynman's famous sum over all possible trajectories, or histories, approach. In quantum mechanics, we are to imagine a given possible pathway of the photon from the photon gun through the screen with the two slits to the photon-counting surface. For each pathway, there is a well-defined procedure to assign an "action." This action can be thought of as having an amplitude and a phase, and the phase rotates through a full circle, 2 pi, many times along the pathway. According to the Feynman scheme, classical trajectories correspond to quantum pathways possessing minimal action.

Consider, says Feynman, all the pathways that start at the photon gun and end up at the same point on the photon-counting surface. Nearly parallel, nearly straight-line pathways, have nearly the same action. So when those pathways interact, they have nearly the same phase, and their interaction yields constructive interference, which tends to build up amplitude. Thus, pathways that are near the classical pathway interact constructively to build up amplitude. By contrast, quirky crooked pathways between the photon gun and the same point on the counter screen have very different actions, hence very different phases, and interact destructively, so their amplitudes tend to cancel. The classical pathway, therefore, is simultaneously the most probable pathway over the sum of histories of all possible pathways, and the pathway that requires the least action.

The result is beautiful, but has two problems. First, Feynman assumes a continuous background space and time in his theory. Quantum gravity, as we will see, cannot make that assumption in the first place. Rather space, or geometry, is a discrete, self-constructing object on its own. Thus, achieving a smooth space and time is supposed to be a consequence of an adequate theory of quantum gravity. If Feynman's sum over histories must assume a smooth background space and time, then it cannot as such be taken as primitive in quantum gravity. Second, granting a continuous background space and time, Feynman's sum over all histories still only gives a maximum of the amplitude for the photon to travel the classical pathway, it never gives an actual photon arriving at the counting surface. No more than any other does Feynman overcome the fundamental linearity of quantum mechanics. We still have to collapse the wave function. Despite these problems, Feynman's results are brilliant, and at least we see a link between the classical and quantum worlds, if not yet actual photons striking counters.

But there is an alternative approach to the link between the quantum and the classical worlds. This possible approach is based on the well-established phenomenon of "decoherence." Decoherence arises when a quantum-coherent Schrödinger wave is propagating and the quantum system interacts with another quantum system having many coupled variables, or degrees of freedom. The consequence can be that the Schrödinger wave function of the first quantum system becomes entangled in very complex ways with the other complex quantum system, which may be thought of as the environment. Rather like water waves swirling into tiny granular nooks and crannies along a rugged fractal beach, the initial coherent Schrödinger equation representing the initial quantum system swirls into tiny and highly diverse patterns of interaction with the quantum system representing the environment. The consequence of this intermixing is decoherence.

To understand the core of decoherence, one must understand that the exhibition of interference phenomena, the hallmark of quantum mechanics noted in the double-slit photon experiment, requires that literally *all* the propagating possible pathways in Feynman's sum over histories that are to arrive at each point on the photon-counter surface, do in fact arrive at that point. If some fail to arrive, the sum over all histories fails. In effect, if some of the phase information, the core of constructive and destructive interference, has been lost in the maze of interactions of the quantum system with its environment, then that phase information cannot come to be reassembled to give rise to quantum interference.

Decoherence is accepted by most physicists. For example, in attempts to build quantum computers that can carry out more than one calculation simultaneously due to the linear features of quantum mechanics, actual decoherence is currently a technical hurdle in obtaining complex quantum calculations.

Decoherence, then, affords a way that phase information can be lost, thereby collapsing the wave function in a nonmysterious fashion. Thus, some physicists hope that decoherence provides a natural link between the quantum and classical realms. Notable among these physicists are James Hartle and Murray Gell-Mann, whose views can be found in Gell-Mann's *The Quark and the Jaguar*. In essence, Hartle and Gell-Mann ask us to consider "the quantum state of the universe" and all possible quantum histories of the universe from its initial state. Some of these histories of the universe may happen to decohere. Hartle and Gell-Mann argue that the decoherent histories of the universe, where true probabilities can be assigned, rather than mere amplitudes, correspond to the classical realm. Others have argued that decoherence itself can be insufficient for classical behavior.

It is striking that there appear to be two such separate accounts of the relation between the quantum and classical worlds, Feynman's sum over histories in a smooth background space-time and decoherence. For an outsider, it is hard to believe that both can be correct unless there is some way to derive one from the other. I will explore one such possibility below. In particular, I will explore the possibility that decoherence of quantum geometries is primary and might yield a smooth space-time in which Feynman's account is secondarily correct.

An Outsider's Doubts

I turn next to an outsider's grounds for doubts about some of the core propositions of quantum mechanics. Roland Omnes, in *The Interpretation of Quantum Mechanics*, is at pains to argue that decoherence is the plausible route to classicity. In his discussion, two major points leap to attention. The first concerns the concept of an elementary predicate in quantum mechanics. Quantum mechanics is stated in the framework of Hilbert spaces, which are finite or infinite-dimensional complex spaces, that is, spaces comprised of finite or infinite vectors of complex numbers. In effect, an elementary predicate is a measurement about an "observable" that returns a value drawn from some set of possible values. And so the first striking point is Omnes' claim that all possible observables can be stated in Hilbert space. The second striking point is Omnes' claim that some observables cannot be observed.

The first point is striking because it is not at all clear that all possible observables can be finitely stated in Hilbert space. My issue here is precisely the same as my issue with whether or not the configuration space of the biosphere is finitely prestatable. As I argued above, there does not seem to be a finite prestatement of all possible causal consequences of parts of organisms that may turn out to be useful adaptations in our or any biosphere, which arise by exaptation and are incorporated in the ongoing unfolding exploration of the adjacent possible by a biosphere.

In quantum mechanics, an observable corresponds to a mathematical operator that "projects out" the subspace of Hilbert space corresponding to the desired observable in a classical measurement context that allows detection of the presence or absence of the observable. But the biosphere is part of the physical universe, and the exapted wings of Gertrude the flying squirrel are manifestly observables, albeit classical observables. If we cannot finitely prestate the observable, "Gertrude's wings," then we cannot finitely prestate an operator on Hilbert space to detect the presence or absence of Gertrude's wings. In short, there seems to be no way to prespecify either the quantum or classical variables that will become relevant to the physical evolution of the universe.

It turns out that the above issues may bear on the problem of time in general relativity, as Lee Smolin realized from our conversations and as I return to shortly.

Now, the second point. Omnes follows up on it. An observable requires a measuring device. There are some conceivable observables for which the measuring device would be so massive that it would, of itself, cause the formation of a black hole. Thus, no information resulting from the measurement could be communicated to the outside world beyond the black hole.

A strange situation; even if we could finitely prestate all possible observables, only some observables can manage to get themselves observed in the physical universe.

What shall we make of conceivable observables that cannot, in principle, be observed? More important, it seems to this outsider, is the following: If observation happens by coupling a quantum system to some other system, quantum or classical, whereby decoherence occurs and, in turn, classicity arises because loss of phase information precludes later reassembly of all the phase information to yield quantum interference, then there appears to be a relation between an observable being observed and the very decoherence by which something actual arises from the quantum amplitude haze.

If that is correct, then only those observables that can get themselves observed can, in principle, become actual. More, it begins to seem imperative to consider the specific possible *pairs* of quantum systems that can couple and decohere, for only thereby can such pairs become classical via decoherence. This begins to suggest preferred histories of the universe concerning such comeasuring pairs of quantum systems. Preferentially, those comeasuring pairs of quantum systems that decohere and become classical will tend to accumulate, due to the irreversibility of classicity. Thereafter quantum-classical pairs of systems that cause decoherence of the quantum system will preferentially accumulate into classicity.

If comeasuring yields classicity, and classicity is irreversible, the classical universe begins to appear to coconstruct itself. In particular, it is generally accepted that bigger systems, that is, systems with more coupled degrees of freedom, *decohere more rapidly when they interact* than smaller systems. If so, this begins to refine the suggestion of preferred histories of the universe concerning comeasuring pairs of quantum systems toward a preference for the emergence of classical diversity and complexity: If quantum systems with more coupled degrees of freedom irreversibly decohere more rapidly into classical behavior when they interact than smaller, simpler systems, then the kinetics of decoherence should persistently favor the irreversible accumulation of bigger, more complex quantum systems, rather than of smaller, simpler, quantum systems.

Chemistry should be an example. Molecules are quantum objects, yet flow into the chemical adjacent possible. The adjacent possible explodes ever more rapidly as molecular diversity, and hence molecular complexity, increases. Reactions of complex molecules are precise examples of the couplings of quantum systems whereby decoherence can happen. Decoherence presumably happens more rapidly among complex reacting molecules than among very simple molecules or the same total number of mere atoms, nucleons, and electrons in the same total volume. This hypothesis ought to be open to experimental test. If confirmed, the flow of possible quantum events into the chemical adjacent possible should, in part, be made irreversible by the decoherence of complex molecular species as they couple and react with one another.

If the general property obtains that complex quantum entities can couple to and interact with other complex quantum entities in more ways than can simple systems and that the number of ways of coupling explodes faster than the diversity of entities, and thus faster than the complexity of those quantum objects, then decoherence should tend to lead to favored pathways toward the accumulation of complex classical entities and processes. I return to these themes below.

The Problem of Time in General Relativity

To my delight, I soon found myself coauthor on a paper with Lee Smolin concerning the problem of time in general relativity. Lee had done the majority of the work, but had taken very seriously my concern that one cannot finitely prestate the configuration space of a biosphere.

In general relativity, space-time replaces space plus time. A history becomes a "world-line" in space-time. But that world-line is a geometrical object in space-time. Time itself seems to disappear in general relativity, to be replaced by the geometrical world-line object in space-time.

But argued Lee, with my name appended, general relativity assumes that one can prestate the configuration space of a universe. In that prestated configuration space, a world-line is, indeed, merely a geometrical object. What if one cannot prestate the configuration space of the universe? If so, one cannot get started on Einstein's enterprise, even if general relativity is otherwise correct. As concrete examples, Lee pointed out that four-dimensional manifolds are not classifiable.

How might one do physics without prestating the configuration space of the universe? Lee postulated use of spin networks, as described below, with the universe constructing itself from some initial spin network. In this picture, time and its passage is real. If there can be a framework in which time enters naturally, and possibly there is a natural flow of time, or an arrow of time preferentially from past to future, then, among other possible consequences, we may be able to break the matter-antimatter symmetry, for antimatter can be stated as the corresponding matter flowing backward in time. Break the symmetry of time in fundamental physics and you may buy for free the breaking of the symmetry between matter and antimatter. If time flows preferentially from past to future, matter dominates antimatter. That would be convenient since matter does dominate antimatter, and no one knows just why.

We will head in this direction.

Spin Networks

For the sixty years following 1926 and the emergence of matrix mechanics and the Schrödinger formulation of quantum mechanics, scant progress was made on quantum gravity. Now, in the past decade or so, there are two alternative approaches, string theory and spin networks. Of the two, string theory has captured the greatest attention. I discuss it briefly below.

Spin networks were invented by Roger Penrose three decades ago as a framework to think about a quantized geometry. Quite astonishingly, spin networks appear to have emerged from a direct attempt to quantize general relativity by Carlo Rovelli and Lee Smolin. In outline, part of the tension between quantum mechanics and general relativity lies in the very linearity of quantum mechanics and the deep nonlinearity of general relativity.

Building on work of Astekar and his colleagues, Rovelli and Smolin proceeded directly from general relativity along somewhat familiar pathways of canonical quantization. In outline, general relativity is based on a metric tensor concerning space-time. The metric tensor is a 4 x 4 symmetric tensor. It turns out that this tensor yields seven constraint equations. The solutions of six of the seven have turned out to be spin networks. The solution of the seventh equation would yield the Hamiltonian function, hence the temporal unfolding, of spin networks in a space x time quantum gravity.

Spin network theories can be constructed in different dimensions. The two most familiar are for two spatial and one temporal or three spatial and one temporal dimension. We will concern ourselves with three plus one spin networks for concreteness. The minimal objects in a spin network are discrete combinatorial objects that constitute first a tetrahedron, with four vertices and four triangular faces. A tetrahedron represents a primitive discrete unit of geometry, or space. Integer-valued labels are present on the edges and vertices of these tetrahedra. The labels on the edges represent spin states. The labels on the vertices represent "intertwinors" and concern how edges entering a vertex are connected to one another into and out of the vertex.

Analytic work has associated an area with a face of a tetrahedron and a volume with its volume. There is, at present, no way to represent the length of an edge connecting vertices. On the other hand, one can think of the integer values on the edges around a face of a tetrahedron as associated with the area of the tetrahedron, such that larger integers correspond to larger areas.

A geometry is built up by minimal moves, called "Pachner moves," in which a given tetrahedron can give rise to daughter tetrahedra off each face. In addition, several tetrahedra can collapse to a single tetrahedron.

Thus we may picture an initial spin network, say, a single tetrahedron. In analogy with chemistry and combinatorial objects, the founder set of a chemical reaction graph, and the adjacent possible in the chemical reaction graph, we may consider the single initial tetrahedron as a founder set, gamma 0. Consider next all possible adjacent spin networks constructible in any single Pachner move. Let these first adjacent possible spin networks lie in an adjacent ring, gamma 1. In turn, consider all the spin networks constructible for the first time from the founder set in two Pachner moves, hence constructible for the first time in one Pachner move from the gamma-1 set of spin networks. Let this new set be the gamma-2 set of spin networks.

By iteration, we can construct a graph connecting the founder spin network with its 1-Pachner move "descendants," 2-Pachner move descendints, . . . N-Pachner move descendents.

Each spin network in each gamma ring represents a specific geometry, subject to the constraint that two spin network tetrahedra that share one triangular face must assign the same spin labels to the common edges, hence, the same area to the common face.

Changes in the values of spins on the edges that change the areas and volumes of the tetrahedra can be thought of as deforming the geometry so that it warps in different ways. However, it should be stressed that there is no continuous background space or space-time in this discrete picture. Geometry is nothing but a spin network, and a change in geometry is nothing but a change in the tetrahedral structure of the spin network by adding or deleting tetrahedra or by changing the spin values on the edges of tetrahedra.

Within quantum mechanics, there is an appropriate way to consider the discrete analogue of Schrödinger's equation, namely a means over time of evolving amplitudes from an initial distribution. In particular, the appropriate means of evolving amplitudes concern what are called "fundamental amplitudes," which specify initial and final values of the integer values on edges before and after Pachner moves.

Consider a given graph linking spin networks from an initial tetrahedron in gamma 0, outward as in a mandala, to all daughter networks in gamma 1, gamma 2, ... gamma N, where N can grow large without limit.

The Emergence of a Large-Scale Classical Limit?

I now describe one approach to thinking about quantum gravity and the emergence of a smooth large-scale geometry based on this mandala and on Feynman's idea of a sum over all histories. Endow the spin networks throughout with the same fundamental amplitudes, thus, the same law propagating amplitudes applies everywhere in the spin network mandala. Begin with all amplitude concentrated in the initial spin network tetrahedron in gamma 0. In this vision, a unit of time elapsing is associated with a Pachner move, such as a move from gamma 0 to a point in gamma 1. With analogy to Feynman's sum over all possible histories, consider the set of all pathways that begin at the initial tetrahedron in gamma 0 and end on a given specific spin network N time steps later, for $N = 1000$. That final spin network might lie in the gamma-0 ring, the gamma-1 ring, the gamma-2 ring, or any ring out to the gamma-N ring.

Here is a hopeful intuition that may prove true. If we consider the family of all histories beginning on gamma 0 and ending in a specific spin network in the gamma $N = 1000$ ring, those pathways must be very similar and few in number. By contrast, if we consider all pathways length 1000 that begin on the gamma-0 tetrahedron and end, 1000 steps later, on a specific spin network in the gamma-23 ring after wandering all over the spin network mandala graph, there may be many such pathways, and they can be very dissimilar. Now, during the amplitude propagation along any pathway, an action can be associated with each Pachner move, hence, we can, with Feynman, think about the constructive or destructive interference among the family of pathways 1000 steps long that begin on the gamma-0 tetrahedron and end on any specific spin network. Then the hopeful intuition is that those pathways that begin on gamma o and end on a spin network member of the gamma $N = 1000$ ring in 1000 Pachner moves will have very nearly the same action, hence, show strong constructive interference. By contrast, those pathways that begin on the gamma-o tetrahedron and end, 1000 Pachner moves later, on a specific spin network in the gamma-23 ring will have very different actions, hence, show strongly destructive interference.

If the constructive interference among the few pathways to ring N overwhelms any residual constructive interference in the inner rings—such as ring 23, due to the larger number of pathways from gamma o to gamma 23—then the hopeful concept is that amplitude will tend to accumulate in the gamma-N ring. Then (goes the hope shared with Smolin) the neighboring spin networks in the gamma-N shell constitute nearly the same geometry and nearly the same action in the sum of histories reaching them, which begins to suggest that a smooth large-scale geometry might emerge.

For this line of theory to succeed, it is not actually necessary that amplitude preferentially accumulate in the outermost, gamma-N, ring. Rather it is necessary as N increases that there be some ring, M, where M is less than N but increases monotonically with N, such that a sufficiently large number of alternative pathways with sufficiently similar phase end on members of the M ring that constructive interference is maximum for members of the M ring. Further, it is necessary that as N increases and M increases, amplitude continue to accumulate on the Mth ring.

In short, the concept is that, via constructive and destructive interference as amplitudes propagate in the mandala, some large-scale smooth geometry will pile up amplitude, hence probability, and a smooth classical geometry will emerge. Here is at least one image of how a large-scale smooth geometry might emerge from spin networks and constructive interference.

At least three major caveats are required. First, no calculation has yet been carried out for such a model, so such a theory may not work. Second, Feynman's sum over histories assumes a classical continuous space and time. It may be entirely invalid to attempt to use a sum over histories argument in this quantum geometry setting. Third, assuming we can use Feynman's sum over histories, we still have possible quantum geometries, not an actual geometry.

Self-Selection of the Laws and Constants of Nature?

Recall the puzzle, nay, the deep mystery, about what processes, if any, might have "chosen" the twenty constants in the standard model such that the universe happens, improbably, to be complex. To answer this deep mystery we have, at present, the anthropic principle, Lee Smolin's concept of cosmic natural selection for black hole density, and the hope to find the ultimate parameter-free theory that would not require multiple universes or a historical process.

With caveats, I now briefly describe a way that may be useful to begin to think about the emergence of the constants such that any universe would have a given set of constants.

Tuning the constants corresponds to tuning the laws of physics. Is there a way to imagine a self-tuning of a universe to pick the appropriate values of its constants, to tune its own laws? I think the answer may be yes. And if the following is wrong in detail, the pattern of thought may prove useful.

In the spin network mandala picture, a 15J symbol, present throughout the spin networks in the mandala, generates an analogue of Schrödinger's equation, hence, the means to propagate amplitudes in the graph of spin networks. Thus, a change in a 15J symbol would correspond to changing the laws of physics about how amplitudes propagate.

Importantly, the fundamental amplitudes are an ordered listing of 15 integers, hence, there is a family of all possible fundamental amplitudes. Since each fundamental amplitude can be thought of as the "law" about propagating amplitudes among spin networks, Louis Crane pointed out that there is an infinite family of all possible laws to propagate amplitude among spin networks.

Thus, imagine an infinite stack of our spin network mandalas, in which each manadala is a graph from gamma 0, the tetrahedron, outward to gamma N, for N allowed to be arbitrarily large, of spin networks reachable in N steps by Pachner moves. The mandala members of the infinite stack of mandalas differ from one another only in the fundamental amplitudes, hence laws, that apply to each mandala. (I concede it may be necessary to have a means to encode in each manadala the given fundamental amplitudes that apply to that manadala.)

Now consider how amplitudes propagate in each mandala from an initial state with all the amplitudes concentrated in the gamma-0 tetrahedron. And consider any two mandalas whose fundamental amplitudes are minimally different. For some such adjacent mandalas with adjacent laws, the small change in the law may lead to a large change in how amplitudes propagate in the mandalas. For other pairs with minimal changes in the fundamental amplitudes or law, the way the amplitude propagates throughout the mandala may be very slight. Assuming this is true, one intuitively imagines that the total system is spontaneously drawn to those tuned values of the fundamental amplitude laws, where small changes in the laws make minimal changes in how amplitudes propagate.

A simple possible mechanism might accomplish this. Imagine a sum of histories from an initial gamma-0 tetrahedron in a mandala with some given fundamental amplitude laws (thereby the initial and boundary conditions are specified), where the pathways in that set of histories pass up and down the stack of mandalas such that the fundamental amplitude *laws* change, as does the spin network, and then consider the bundle of all such histories that end on a given spin network in a given gamma ring with given, perhaps new, fundamental amplitude laws. In effect, this conceptual move allows there to be quantum uncertainty not only with respect to spin networks along histories, but also quantum uncertainty with respect to the *law* by which amplitude propagates.

Then one can imagine a sum over all histories that, by constructive interference alone, picks those pathways, hence fundamental amplitude laws, that minimize the change in the ways amplitudes propagate. Such pathways would have similar phase, hence, accumulate amplitude by constructive interference. Then, by mere constructive interference, one can hope that such a process would pick out not only the history, but also tune the law to the well-chosen fundamental amplitudes laws that maximized constructive interference. Hopefully, that constructive interference would pick out smooth large-scale geometries like classical flat or near flat space. In such a large-scale classical-like space and time, Feynman's familiar sum over histories that minimizes a least action along classical trajectories would emerge as a consequence.

Smolin and I discuss this possibility in a second paper. I find the idea attractive as a research program because it offers a way in which a known process, constructive interference, modified to act over a space of geometries and laws simultaneously, chooses the law. It is, of course, rather radical to suppose that there is quantum uncertainty in the law, but it does not seem obviously impossible.

On an even grander scale, particle physicists build the standard model from an abstract algebra called $SU(3)$ x $SU(2)$ x $U(1)$. One can imagine a similar research program that by constructive interference alone picks out the particles, constants, and laws of the standard model. Presumably, particles governed by sufficiently "nearby" laws would be able to interact, hence undergo constructive or destructive interference, thus picking the particles and the laws simultaneously.

There is a further interesting feature, for we appear to have in our mandala, or mandalas, a new arrow of time. Allow that at any step, any Pachner move can happen. Some moves add tetrahedra. A equal number delete tetrahedra. Yet the number of spin networks in ring $N + 1$ is larger than the number of spin networks in ring N. Statistically, there are more ways to advance into ring $N + 1$ than to retreat from ring N into ring $N - 1$. Other things equal, amplitude should tend to propagate outward from the gamma-o tetrahedron. There is an analogy in chemical reaction graphs to the adjacent possible and the real chemical potential across the frontier from the actual to the adjacent possible.

But if so, time enters asymmetrically due to the graph structure of the spin network mandala. Then, statistically, time tends to flow in one direction, from simpler toward more complex spin networks into the ever-expanding adjacent possible.

A Brief Comment on String Theory

String theory has gained very substantial attention as a potential "theory of every-thing," namely, a theory that might link all four forces and all the particles of the standard model into a single coherent framework. I do not write with even modest expertise on the subject. Nevertheless, it is possible that the concept of the law selecting itself via maximum constructive interference in a sum over all possible histories in a space of both spin networks and laws might possibly have relevance to string theory. The description of string theory that I give draws heavily on Brian Greene's *The Elegant Universe*.

As is known qualitatively by many outside the confines of the physics community, string theory began by giving up the picture of fundamental particles as zero-dimensional, point particles. In its initial version, in the place of point particles, string theory posited one-dimensional strings that might be open, with two ends, or closed loops, with no free ends. Among the fundamental ideas of string theory is the idea that the different particles and the different forces can all be thought of as different modes of vibration of such strings. Because strings have finite length, string theory can hope to overcome the infinities that emerge when attempts are made to marry point particle quantum theories with general relativity in a contin-uous space-time. In effect, the finite length of the strings prevents consideration of space becoming infinitely curved at a point. Thus, string theory can dream of unit-ing quantum mechanics and general relativity, and it has, in fact, produced the en-tity carrying the gravitational force, the graviton, in a natural way.

Current string theory has gone beyond single-dimensional strings, and now considers two-or-higher-dimensional entities called M-branes. The rough present state of the art has shown that there are at least five one-dimensional string theo-ries and M-brane theory. All of these theories appear to be linked as cousins of one another via various dualities among the theories.

String theories posit either eleven-or-fewer-dimensional space and time, with three of the spatial dimensions unfurled and large scale, corresponding to our fa-miliar three-dimensional space. The remaining dimensions are imagined as curled up on the Planck length scale in what are called "Calabi-Yau" spaces, or more gen-erally, compactified moduli. Compactification of an eleven-dimensional space and time can be thought of as a large-scale three-dimensional space and time, but with the additional dimensions curled up at each point in the large-scale three-dimen-sional space.

Calabi-Yau spaces can have different topologies. Consider as an analogy a long thin tube with two ends and a one-hole torus, like a donut. These two are topolog-ically different. As a consequence, closed one-dimensional string loops can "live" on these surfaces in different ways. Thus, if you think of a string as a closed loop, that loop might live on the long tube in two ways, either wrapped around the tube one or more times or not wrapped around the tube, but lying on the tube's surface like a rubber band lying on a surface. By contrast, consider the torus. The closed string might wrap around the torus in either of two ways, through the hole or around the torus. In addition, the string loop might live on the surface of the torus without wrapping either dimension. Each of these different ways of being on the tube or torus and the corresponding modes of vibration constitute different particles and forces. Calabi-Yau spaces are more complex than the tube or torus, but the basic consequences are the same. Different Calabi-Yau spaces, or more generally, different compactified moduli, with different kinds of holes around which strings can wrap zero, one, or more times correspond to different laws of physics with different particles and forces.

Physicists have shown, furthermore, that one Calabi-Yau space can smoothly deform into another with a "gentle" tearing of space and time. Hence, the laws, forces, and particles can deform into one another in a space of laws, forces, and particles. Within current string theory, it appears that it is still not certain that there exists a Calabi-Yau space whose string or M-brane inhabitants would actually correspond to the known particles and forces, but hopes are high.

However, even if there is a Calabi-Yau space whose strings and M-branes do correspond to our known particles and forces, string theorists have the difficulty that it is not clear how the current universe happens to choose the correct Calabi-Yau space. The familiar ideas on this subject include the existence of a multiverse and the weak anthropic principle. For example, one could imagine Lee Smolin's arguments for choices of Calabi-Yau spaces that lead to fecund universes with a near maximum of black holes, which are the birthplaces of still further universes.

The parallel between spin networks with different fundamental amplitude laws and the family of string and M-brane theories that can deform into one another is that in both theories we confront a family of theories having the property that different members of the family correspond to different particles, forces, and laws. In both cases, physicists do not at present have a theory to account for how, in this embarrassment of riches, our universe happens to pick the correct laws. I therefore make the suggestion that the same pattern of reasoning that I described above, a sum over histories of trajectories that vary both in configurations and in the laws, which maximizes constructive interference, might prove a useful approach. In the string theory context, one would consider a hyperspace of Calabi-Yau spaces, in which neighboring Calabi-Yau spaces would propagate amplitudes from the same initial condition in different ways. Presumably, somewhere in the hyperspace of Calabi-Yau spaces, small changes in Calabi-Yau spaces would yield small changes in how amplitudes propagate. For other locations in the hyperspace of Calabi-Yau spaces, small changes in the Calabi-Yau space would yield large differences in how amplitudes propagate. In the hyperspace of Calabi-Yau spaces, where one Calabi-Yau space can deform into its neighbors, it should be possible to construct a sum over all histories of trajectories between an initial and final state in the same or different Calabi-Yau space, then seek such sums over histories that maximize constructive interference. The hope is that maximizing constructive interferences would pick out the Calabi-Yau space corresponding to our particles and forces. Presumably, this would occur in the region of the hyperspace of Calabi-Yau spaces, where small changes in the Calabi-Yau space yield the smallest changes in how amplitudes propagate. In short, maximization of constructive interference may be a useful principle to consider to understand how the universe chooses its laws.

String theorists recognize the need to confront further major problems. Most notably, string theory posits a background space-time in which strings and *M*-branes vibrate. But if string theory is to be *the* theory of everything, including space and time, then space and time cannot be assumed as a backdrop. Thus, a virtue of spin networks is that it affords the hope of a quantized geometry from the outset. On the other hand, particles and the three nongravitational forces have yet to be incorporated into a spin network picture.

A Decoherence Spin Network Approach to Quantum Gravity

However the universe picks its presumably quantum laws, somehow the classical realm emerges. I noted above that current theory sees two approaches to linking the quantum and classical realms. The first is based on Feynman's sum over histories, but as a perturbative theory assumes a continuous background space-time and does not get rid of the linear superposition of possibilities that is the core of quantum mechanics and interference.

What of the second approach, decoherence? The reality of decoherence is established. If one is to take decoherence seriously, and also to consider geometry constructing itself, then presumably decoherence can apply to geometry as it constructs itself. What would it mean to apply decoherence to quantum gravity itself, to the vacuum, to geometry itself?

Well, to an outsider, the following seems possible. If we are to conceive of an initial spin network, say a tetrahedron, and all possible daughter spin networks, as well as all their possible daughter spin networks, propagating amplitudes on the mandala, then at any moment N steps away from moment 0, more than one geometry is possible—namely all those reachable in N Pachner moves.

We seem to confront the same problem we confront with quantum systems coupling to quantum systems, such as electrons coupling to organic molecules, or to classical systems, such as rocks. These quantum systems can decohere. Can quantum geometries become coupled with one another or different parts of one quantum geometry become coupled, so to speak, and decohere?

Why not try the idea?

I now discuss one possible approach to this issue. The approach posits a quantum of action, h, to the generation of a tetrahedron, hence a Planck energy and thus a Planck mass to a tetrahedron, and decoherence setting in at a sufficient mass and size scale.

By use of an equation suggested by Zurek relating the decoherence timescale, Td, to the relaxation timescale, Tr, of the system, in which increasing mass and area increase the rate of decoherence in proportion to their product, it can be qualitatively shown (via sufficiently rough arguments) that geometry may well be thought of as decohering, and doing so on a length scale of about 10^{-15} cm, which is smaller than the Compton radius of the electron and even smaller than the radius of a nucleus.

Now, there are some interesting features of this rough calculation. First, if we begin with an initial tetrahedron of geometry, it can have four daughter tetrahedra. In turn, each daughter tetrahedron can have two or more daughter tetrahedra, hence, the initial spin network can grow *exponentially* in the number of tetrahedra before decoherence sets in. This is a clue that a purely quantum account might be given of an initial exponential expansion of a universe starting with a single tetrahedron. Thus, it might be possible to do without the "inflationary hypothesis" of exponential expansion of a classical space in the early moments after the big bang.

Second, an initial exponential expansion of geometry might overcome, as the inflationary hypothesis does, the particle-horizon problem in cosmology, in which we confront the puzzle of why parts of the universe that have been out of apparent causal contact since the big bang can be so similar. If the initial expansion is exponential, then slows to linear, as in the inflationary hypothesis or perhaps in this purely quantum approach, then the particle-horizon problem may disappear.

Third, a purely quantum exponential expansion over many orders of magnitude should presumably yield a flat power spectrum over many size scales for quantum fluctuations in the earliest small fraction of a second of the life of the universe prior to the end of this exponential expansion when decoherence of geometry occurs.

Fourth, we must consider when geometries decohere whether there may be geometries that are the slowest to decohere. If different parts of a single spin network geometry can become coupled, it is natural to assume that flat parts might decohere more slowly than distorted parts of that geometry. Intuitively, phase information can get lost more readily when two lumpy parts of a geometry couple than when two flat parts of a geometry couple. Of course, an explicit model exploring this is badly needed and entirely missing, but in its absence, let's make the assumption. Given that, then an initial exponential explosion of flat and warped geometry occurs until decoherence sets in on a length scale of something like 10^{-15} cm. At this point, flat geometry "wins" because it decoheres most slowly. Hence, as soon as decoherence of geometry sets in, space tends to be flat in the absence of matter.

But even after decoherence sets in, geometry is busy all the time trying to build geometry exponentially and everywhere, while simultaneously decohering. Now an interesting feature of the Td/Tr equation alluded to above is that whatever the exponential rate of expansion of geometry may be per Planck time unit, the exponential rate of decoherence, Td, which grows as the mass times the size scale squared of the geometry, increases, until eventually the exponential rate of formation and exponential rate of decoherence of geometry must balance. The exponential expansion of the universe is over. However, linear expansion by construction of geometry can continue. The fastest linear construction of geometry from any tetrahedron would be at the speed of light.

When the rate of geometry formation and decoherence balance, geometry keeps building tetrahedra as fast as possible everywhere, but flat geometry, by hypothesis, decoheres most slowly. In the limit, perhaps flat geometries do not decohere at all. Then, the geometry of the universe tends to be flat in the absence of matter, as Einstein requires in general relativity. And, once again, the flatness after exponential expansion may overcome a need for the inflationary scenario and solve the particle-horizon problem.

It may be of interest that the assumption of an action, h, in the generation of each tetrahedron implies an expanding total energy in geometry itself, the vacuum itself, as geometry constructs itself. Indeed, one expects that the assumption of an action, h, per tetrahedron would lead to a uniform energy density, a constant scalar quantity even as geometry grows. Such an energy could be related to the cosmological constant.

It may also be of interest that string theory posits "extra dimensions" that "curl up" on themselves to yield four-dimensional space-time. Could the decoherence of geometry afford a parallel way that extra dimensions can curl up? And could the ever-generating possible geometries, as they generate exponentially even as they decohere, yield sufficient extra degrees of freedom to correspond to the modes of oscillation of strings or M-branes in six or seven extra dimensions?

It may be interesting that the energy content of geometry could be enormous compared to that of familiar particles of the same size scale. That might allow the familiar particles with rest mass to borrow a tiny bit of the vacuum energy for their mass. Such a possibility could hint that matter, energy, and geometry might be able to interconvert. Perhaps different particles would be different kinds of topological "knots" in the spin network structure of geometry with interconvertible particles being nearby knot topologies.

We are obviously far from anything like a coherent theory that implements any of the intuitions above. They remain at best mere suggestions for a research program.

Coconstructing Complexity?

I began this chapter wondering why the universe is complex. In place of the anthropic principle or Lee Smolin's cosmic selection, I have suggested one possible approach to the choices of the constants of nature by maximizing constructive interference over a sum of all histories through a space of both configurations and laws. Even if that program were to succeed, it does not necessarily yield a complex universe, let alone one poised roughly between expansion and contraction.

Might we see ways to understand why the universe is complex? Perhaps, but merely perhaps. I return to the thoughts earlier in this chapter that decoherence requires coupling systems and the loss of phase information. If, in general, complex and high-diversity quantum systems with many coupled degrees of freedom lose phase information when they interact more rapidly than an equal number of simple, low-diversity quantum systems with the same total number of interacting parts, then the comeasuring of entangled quantum systems should tend toward higher complexity, diversity, and classicity. In short, complexity and diversity would beget classicity irreversibly. In turn, this would lead to a preferred tendency toward a lock-in of complexity and diversity. There is a sense in which classical objects are like the constraints on the release of energy that permits work to be done. Classical objects, interacting with quantum objects, lead to decoherence and more classicity. Complex pairs of quantum objects that decohere readily, or classical objects and those quantum objects that are caused to decohere readily when interacting with the classical object, form preferred pairs that tend to decohere, hence become frozen into classicity. We begin to have an image of preferred pairs of quantum systems coupling and decohering, hence, an image of a complex and diverse universe constructing itself as it nonergodically invades the adjacent possible, rather as a biosphere constructs itself. And if more complexity and diversity means more comeasurement and faster decoherence of a wider variety of complex quantum systems, in analogy with the concept that extracting work from increasingly subtle nonequilibrium systems requires increasingly subtle measuring and coupling devices, the universe as a whole may persistently break symmetries as new entities come into existence, and hence expand its diversity, complexity, and classicity as fast as possible.

Loose arguments? Yes. Testable? Here and there. Wrong? Probably. Deeply wrong? Maybe not. Does this get the universe to the edge of expansion versus contraction or to a flat universe expanding forever more rapidly? I would love it to be so. Indeed, I would love a view in which matter, energy, and geometry can all interconvert. After all, if geometry, the vacuum, has energy, such interconversion does not seem impossible. Do the considerations of this chapter require detailed models and supporting calculations to be taken as more than the merest suggestions? Absolutely. This chapter, like much of *Investigations*, is protoscience. But science grows from serious protoscience, and I take *Investigations* to be serious protoscience.

We enter a new millennium. There will be time for new science to grow.

EPILOGUE

I CLOSE *Investigations* by returning to the puzzle of reductionism. In Stephen Weinberg's book *Dreams of a Final Theory*, he makes a strong case for a kind of unity of science in which he argues that our answers to the question "why?" always and ultimately lead us to the fundamental particles and forces. In saying this, Weinberg is fully sympathetic to our practical incapacity to predict, to quantum uncertainty, to chaos theory, to Phil Anderson's "more is different" statements on emergence, and to the intermarriage of law and history.

But I wonder if Weinberg is right. Is it really true that the answers to "why?" always lead us downward to the hoped-for ultimate particles and forces? To make the points I want to make, I ask you to grant me some conceptual possibilities. Let us return to string theory and the Calabi-Yau spaces that string theorists are considering. As noted in the previous chapter, different Calabi-Yau spaces would yield different particles and forces. Suppose that there actually is a Calabi-Yau space whose particles and forces correspond to our own. And suppose that there are nearby Calabi-Yau spaces whose particles and forces are close to our own. Further, suppose that for those nearby Calabi-Yau spaces, the resulting particles and forces would still yield nucleons, atoms, chemical reactions, and molecules, but that the resulting atoms, nuclear and chemical reactions, and molecules might be somewhat different from our own.

With our current particles and forces, nucleosynthesis of elements heavier than hydrogen, helium, and lithium occurs in stars and supernovae via an autocatalytic cycle called the Bethe cycle, involving carbon, nitrogen, and oxygen. That is, the current particles, nuclei, and nuclear reactions that transmute the elements happen to allow this specific autocatalytic cycle to form. Suppose the particles, forces, and laws were slightly different, with slightly different nuclear reactions, and that once again, some autocatalytic nuclear reaction cycle were to form. More generally, conceive of a theory that asks, for a family of Calabi-Yau spaces and a family of particles, forces, nucleons, and nuclear reactions, what the statistical features of networks of nuclear reactions are and what the probabilities over such a space of particles and forces are that autocatalytic nuclear reaction cycles arise. Now suppose that we were to find that it is utterly typical for any one of the set of laws, particles, and forces that one or more such autocatalytic nuclear reaction cycles would form such that light elements are built up to complex atomic nuclei.

And suppose that for any of the resulting different chemistries, complex chemical reaction networks would be possible under the corresponding quantum chemistry given the set of laws for that Calabi-Yau space. Further suppose that the statistics of such chemical reaction networks were such that at a sufficient chemical diversity, autocatalytic chemical systems were very likely to arise. And further suppose that for the entire family of such Calabi-Yau spaces, the autocatalytic sets that would be expected to arise might also link exergonic and endergonic reactions to create work cycles. In all these cases, with a family of different particles, forces, and laws, we could still get complex molecules and autonomous agents that would evolve by Darwinian natural selection to build biospheres.

What would we say were we to find that these results are all possible within some *family members* of Calabi-Yau spaces? What happens to our impulse to say that we have "reduced" autonomous agents to chemistry thence to physics and to the fundamental particles, forces, and laws? My own feeling is that we would say something quite different, for we would have seen that any one member of the family of Calabi-Yau spaces is sufficient, but not necessary and sufficient, for the emergence of autonomous agents and biospheres.

Indeed, we begin to glimpse a *constructivist* companion to the reductionist thesis. The expected mathematical features of nuclear reaction networks and chemical reaction networks that engender autocatalytic nuclear systems and chemical reaction systems, the expected closures of work and catalytic tasks that autonomous agents achieve, the propagating organization that autonomous agents coconstruct, and the nonergodic exploration of the vast adjacent possible for the universe and an ecosystem begin to seem to have as much legitimacy as law—and fundamental law at that—as do the reductionist successes we so rightly honor.

Sometime in the next half century we will, almost certainly, begin to study the coconstruction of tiny biospheres carried out by molecular autonomous agents that we ourselves shall have created. We will witness the propagating diversifying organization, the unexpected adaptations. We will search for constructivist laws true of any biosphere. We will found a general biology. And we will be spellbound.

Investigations has been but one strand in the beginnings of such a constructivist science. I pray further work will weave more strands into a rich new tapestry.

REFERENCES

Alberts, Bruce, Dennis Bray, Julian Lewis, Martin Raff, Keith Roberts, and James D. Watson. *Molecular Biology of the Cell*. 3rd ed. New York: Garland Publishing, 1994.

Anderson, P. W. "More is different", *Science*, 177: 393–96, 1972.

Bak, Per. How *Nature Works: The Science of Self-Organized Criticality*. New York: Copernicus, 1996.

Davies, Paul C. W. The Fifth Miracle: *The Search for the Origin and Meaning of Life*. New York: Simon & Schuster, 1999.

Dennett, Daniel C. *Darwin's Dangerous Idea: Evolution and the Meanings of Life*. New York: Simon & Schuster, 1995.

Eigen, Manfred, and Peter Schuster. *The Hypercycle: A Principle of Natural Self-Organization*. New York: Springer, 1979.

Gilbert, Scott F. *Developmental Biology*. 2nd ed. Sunderland, MA: Sinauer, 1988.

Greene, Brian. *The Elegant Universe: Superstrings, Hidden Dimensions, and the Quest for the Ultimate Theory*. New York: W. W. Norton, 1999.

Guth, Alan H. *The Inflationary Universe: The Quest for a New Theory of Cosmic Origins*. New York: Addison Wesley, 1997.

Kauffman, Stuart A. *At Home in the Universe: The Search of the Laws of Self-Organization and Complexity*. New York: Oxford University Press, 1995.

———. *Investigations*. Santa Fe Institute preprint. 1996.

———. *Origins of Order: Self-Organization and Selection in Evolution*. New York: Oxford University Press, 1993.

Layzer, David. *Cosmogenesis: The Growth of Order in the Universe*. New York: Oxford University Press, 1990.

Leff, Harvey, and Andrew Rex, eds. *Maxwell's Demon: Entropy, Information, Computing*. Princeton, NJ: Princeton University Press, 1990.

Maddox, John. *What Remains to Be Discovered*. London: Macmillan, 1998.

Monod, Jacques. *Chance and Necessity: An Essay on the Natural Philosophy of Modern Biology*. Trans. Austryn Wainhouse. New York: Knopf, 1971.

Morowitz, Harold J. *The Foundations of Bioenergetics*. New York: Academic Press, 1978.

Omnes, Roland. *The Interpretation of Quantum Mechanics*. Princeton, NJ: Princeton University Press, 1994.

Peebles, P. J. E. *Principles of Physical Cosmology*. Princeton, NJ: Princeton University Press, 1993.

Pimm, Stuart. *The Balance of Nature? Ecological Issues in the Conservation of Species and Communities*. Chicago: University of Chicago Press, 1991.

Prigogine, Ilya. *The End of Certainty: Time, Chaos, and the New Laws of Nature*. New York: Free Press, 1997.

Raup, David. *Extinction: Bad Genes or Bad Luck*. New York: W. W. Norton, 1991.

Rosen, Robert. *Life Itself: A Comprehensive Inquiry into the Nature, Origin, and Fabrication of Life*. New York: Columbia University Press, 1991.

Schrödinger, Erwin. *What Is Life? The Physical Aspect of the Living Cell*. Cambridge: Cambridge University Press, 1944.

Smolin, Lee. *The Life of the Cosmos*. New York: Oxford University Press, 1997.

Weinberg, Stephen. *Dreams of a Final Theory: The Search for the Fundamental Laws of Nature*. New York: Pantheon Books, 1993.

Wittgenstein, Ludwig. *Philosophical Investigations*. 3rd ed. Trans. G. E. M. Anscombe. New York: Prentice Hall, 1973.

INDEX